Cover: Old millstones on a Derbyshire hillside.
Inside cover: The cobbles in this Edinburgh street are made of rock formed in the core of an ancient volcano.
Left: A fern-like growth of pure gold from Devon.

1

ROCK SOLID

Britain is a small island on a medium-sized planet – a planet that was formed, with nine other planets and a Sun, 4600 million years ago from the swirling debris left over after a cluster of stars exploded.

Little is known about the first few hundred million years of the Earth's history. We know that by 4000 million years ago, a solid crust had formed. By 3960 million years ago, there were continents and oceans. Rain fell on the continents and eroded them, washing sand and fragments of rock into the sea. The processes of shaping the surface of the planet had begun, never to stop. But in these early *Hadean* and *Archaean* eras of Earth history, Britain did not exist; it had not been born. The country's oldest rock, from Northwest Scotland, is 2800 million years old; it formed when the Earth was already 1800 million years old, over a third as old as it is now.

The crust is not static – it is ever changing. The continents have shifted their positions about the globe throughout geological time. This process of **continental drift** was ridiculed when it was first mooted in 1920, but it is now an established fact. Quite where our 2800 million-year old chunk of Northwest Scotland was when it was first formed has not been determined, but it is certainly well-travelled. Fig 1 shows the changing positions of the continents since 500 million years ago. During the globe-trotting, continents came together and parted many times in a multitude of configurations. Chunks of crust broke away from continents, and adhered to other chunks. This process has, over hundreds of millions of years, transformed the continents into a patchwork of chunks of crust or **terranes**.

The terranes that make up the foundations of Britain's crust are shown in fig 3. They have been in this configuration since about 320 million years ago, when the Lizard Peninsula terrane in Cornwall 'docked'. Britain has moved about since then, but has not acquired any more terranes to its patchwork. Sea levels have changed, and Britain has spent many millions of years submerged beneath the waves, accumulating sediments, which have since turned to rock, on top of its terranes. On occasions, the terrane patchwork has been punctured by

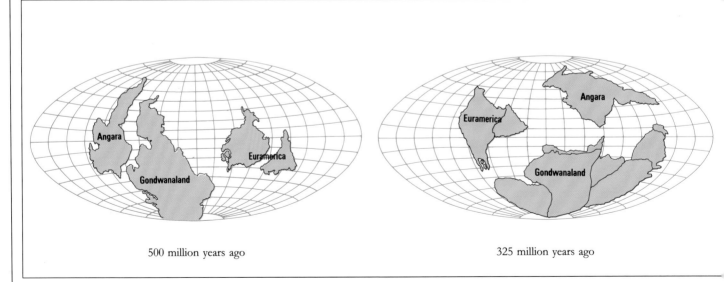

500 million years ago

325 million years ago

1 The drift of the continents over the globe during geological time.

Geologists talk in blasé fashion about millions and billions of years – it is a totally different timescale from the pattern of our daily lives, and involves a completely different way of thinking. The 'New' Red Sandstone, at around 250 million years old, is a comparative youngster in relation to the 4600 million years of the Earth's age. Yet at 249.5 million years before the earliest recorded human history, the New Red Sandstone is unquestionably ancient, and laughably ill-named in the context of our own lifespans of three score years and ten. To give a visual impression of the immensity of geological time, fig 2 shows the events of Earth history compressed into a year: imagine the Earth as having formed at one minute past midnight on the first of January. The first human beings *(Homo habilis)* would have appeared at 8pm on 31 December, and the Earth's New Year's Eve celebrations were in full swing by the time we, *Homo sapiens*, gatecrashed the party.

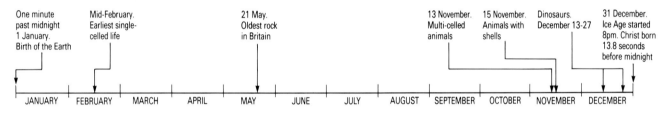

| One minute past midnight 1 January. Birth of the Earth | Mid-February. Earliest single-celled life | | | 21 May. Oldest rock in Britain | | | | 13 November. Multi-celled animals | 15 November. Animals with shells | Dinosaurs. December 13-27 | 31 December. Ice Age started 8pm. Christ born 13.8 seconds before midnight |

| JANUARY | FEBRUARY | MARCH | APRIL | MAY | JUNE | JULY | AUGUST | SEPTEMBER | OCTOBER | NOVEMBER | DECEMBER |

2 The geological timescale compressed into a year.

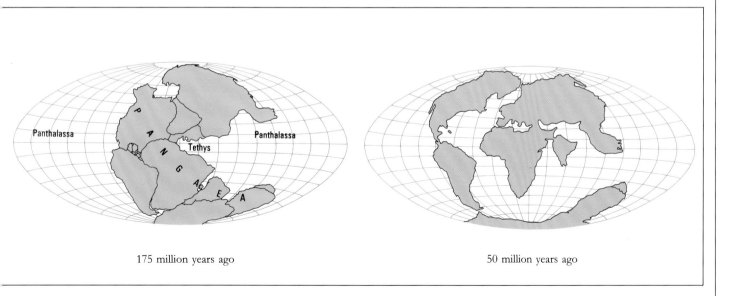

175 million years ago   50 million years ago

ROCK SOLID

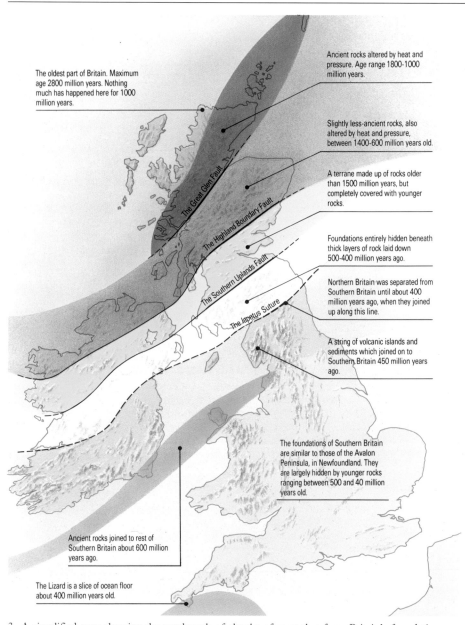

The oldest part of Britain. Maximum age 2800 million years. Nothing much has happened here for 1000 million years.

Ancient rocks altered by heat and pressure. Age range 1800-1000 million years.

Slightly less-ancient rocks, also altered by heat and pressure, between 1400-600 million years old.

A terrane made up of rocks older than 1500 million years, but completely covered with younger rocks.

Foundations entirely hidden beneath thick layers of rock laid down 500-400 million years ago.

Northern Britain was separated from Southern Britain until about 400 million years ago, when they joined up along this line.

A string of volcanic islands and sediments which joined on to Southern Britain 450 million years ago.

The foundations of Southern Britain are similar to those of the Avalon Peninsula, in Newfoundland. They are largely hidden by younger rocks ranging between 500 and 40 million years old.

Ancient rocks joined to rest of Southern Britain about 600 million years ago.

The Lizard is a slice of ocean floor about 400 million years old.

The Great Glen Fault.

The Highland Boundary Fault.

The Southern Uplands Fault.

The Iapetus Suture.

3 A simplified map showing the patchwork of chunks of crust that form Britain's foundations.

volcanoes. The net result is a complicated three-dimensional puzzle which is still being unravelled by geologists, but many terrane boundaries are clearly visible in the landscape, such as the Great Glen fault in Scotland. The variety of rocks in our terranes has given us our rich and varied landscape, our history and industrialised culture.

4 The Great Glen fault, Scotland.

BRITAIN NOW

The pattern of the continents on the globe today is shown in fig 5. The crust is (and always has been) divided into sections or **plates**. 'Seams' where these plates join are also shown – they also mark lines of frequent earthquakes and volcanic activity. None of these present **plate boundaries** run through Britain – we have been in the middle of a plate for at least 320 million years.

The map (fig 7) shows the solid rocks that make up the surface of Britain today (superficial deposits such as river gravels and peat are not shown). The numbers show the ages of the rocks in millions of years. Some of the terranes shown in fig 3 can still be picked out in fig 7, but many of the boundaries are obscured by a cover of younger rocks.

Britain contains a remarkable diversity of types and ages of rock in a comparatively small land-area. Geology is a young science, and geological maps did not start to be drawn until the early 1800s. The picture is continually being updated as new techniques are developed and more detailed surveys are carried out, bringing new information to light.

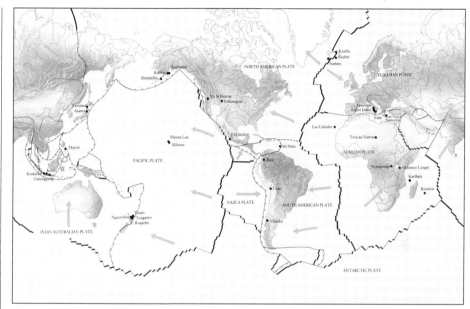

5 The positions and shapes of the Earth's plates today.

6 Britain's oldest rocks are found in the Scourie region of Northwest Scotland.

ROCK SOLID

CENOZOIC (the age of mammals)

Tertiary and Marine Pleistocene: **00 – 65**
Mainly clays and sands.

MESOZOIC (the age of the dinosaurs)

Cretaceous: **65 – 145**
The age of chalk.

Jurassic: **145 – 208**
A mixture of clays, such as the fossil grounds of Lyme Regis and the honey-coloured limestones of the Cotswolds and Lincolnshire.

Triassic: **208 – 240**
Mainly orangey-red sandstones.

PALAEOZOIC (from sea shells to reptiles)

Permian: **240 – 290**
Orangey-red sandstones laid down in a hot desert; and in eastern England, a creamy type of limestone called dolomite, used to build the Houses of Parliament.

Carboniferous: **290 – 360**
The age of limestone and coal.

Devonian: **360 – 408**
Mainly deep red sandstones. Slates in Cornwall and North Devon.

Silurian: **408 – 440**
Shales, sandstones and limestones.

Ordovician: **440 – 510**
Rocks that formed on the bed of an ancient ocean.

Cambrian: **510 – 550**
The oldest rocks in which fossils are common.

Late Precambrian: **600 – 1000**
Sandstones in Northwest Scotland, mainly slates in England and Wales.

Precambrian: **600 – 1800**
The complex metamorphic rocks of the Scottish Highlands.

Early Precambrian: **1500 – 2800**
The ancient rocks of Northwest Scotland.

IGNEOUS ROCKS

Intrusive: **2000 – 50**
Rocks formed from a melt which cooled deep inside the Earth's crust – mostly granite.

Extrusive, or *Volcanic:*   **500 – 50**
Rocks that have been formed from volcanoes.

7 A simplified map showing the rocks that make up the surface of Britain.

EDINBURGH

8 Edinburgh 325 million years ago and (inset) today: castle (foreground) and Arthur's Seat (background), two of Edinburgh's extinct volcanoes.

Edinburgh is a city of contrasts – a leading cultural capital, yet not short on social problems. It has been nicknamed both *The Athens of the North*, because of the elegance of Georgian terraces and public buildings built of the fine locally quarried Craigleith sandstone; and *Auld Reekie* because of the grime, filth and soot that clings to many of the buildings after two hundred years of burning the results of a once-prosperous coal mining industry. Both these nicknames, and all Edinburgh's social, economic and cultural development, have their roots in an age called the **Carboniferous**, that lasted from 345 to 280 million years ago. Then, Britain was at the equator, and the lowlands of Scotland were a series of warm blue lagoons and lush green tropical islands, interrupted by the violence of a multitude of volcanoes.

Perhaps the most obvious contrasts, certainly to the first-time visitor, are in the landscape and skyline. Shooting up from the bustling streets are several steep, rugged hills with almost vertical faces of jagged black rock. Perched on the top of the most central of these is the castle, impenetrable fortifications built into the top of the austere cliffs in such a way that it is hard to see where rock ends and wall begins (see fig 24, p.14). A more strategic vantage point would be hard to imagine. Its permanence and solidity seem unquestionable now, yet these tall crags were once molten lava churning in the throat of a volcano that last erupted 325

ROCK SOLID

9 Samson's Ribs – basalt columns on the slopes of Arthur's Seat.

10 Register House, Edinburgh, built mainly of Craigleith sandstone.

million years ago. The lava cooled and solidified; the volcano has long since been at rest.

The tallest, and most impressive of Edinburgh's volcanic hills is Arthur's Seat which rises 823 feet in Holyrood Park. It could be a wild, heather-clad highland mountain, yet there it stands, bang in the middle of a city – a magnificent site for kite flying, for long rambles, and for studying the inside of an extinct volcano. The summit is another congealed volcanic throat which once spewed out fast-flowing sheets of red-hot lava down its slopes, and filled the air with ash and debris. Because it has been extinct for well over 300 million years, you can quite safely sit down on the now cold, hard, black rock and admire the view.

The layers of ash and lava slope away from the summit of the volcano; a little more steeply now than 325 million years ago, as the land has been tilted by earth movements. Black **basalt**, the commonest type of lava on Earth, shows the characteristic hexagonal columns, like the Giant's Causeway in Ireland, resulting from contraction during cooling from molten lava to solid rock. On a clear day many other remnants of old volcanoes can be seen protruding as hills from the gentle undulations of the Scottish lowlands – the Castle Rock in Edinburgh itself, Berwick Law on the eastern horizon, and the Lomond Hills across the Firth of Forth.

If you could travel back 325 million years and have had the courage to sit on the lip of the boiling crater of Arthur's Seat, you would have seen beneath you forests of giant club mosses and tree-ferns colonising the sands and silts of a river delta. The Highlands would have been much higher than they are today – of Himalayan proportions – and would have been a much more prominent feature of the skyline. The sands and silts washed down from these mountains into the rivers and deltas are now compressed and petrified into the rocks of the Fife and Midlothian Coalfield.

To the west, a large lagoon was populated by sharks and other fish. There were also millions of smaller plants and animals which died, sank to the bottom and, over millions of years, slowly decomposed to form oil. The oil-soaked sediment of the ancient lagoon became the Scottish oil shales, and in the last century oil was extracted from them by a heating and distillation process. The tips, or **bings**, of red-baked shale still litter the countryside around the M8 and M9 motorways. The sea, as today, was not far off, but it was a good deal warmer and would occasionally wash over the land and lap at the foot of the volcanoes.

JAMES HUTTON – THE FATHER OF GEOLOGY

Before the seventeenth century, not many people studied geology or the rocks around them, but those that did took the view that all rocks were laid down in the sea around the time of the Great Flood, and that the surface of the Earth had remained unaltered ever since. Fossils found in some rocks were cited as examples of the poor creatures that were left behind to drown as the Ark set sail. It had been noticed that some rocks were made up of a mesh of crystals, and it was assumed that they had crystallised out of a solution, in the same way that salt will crystallise out of seawater if it is left to evaporate.

11 James Hutton (1726-1797) was variously a lawyer, doctor, chemist and farmer as well as a geologist. Maybe it was this varied experience that enabled him to think laterally and interpret rocks and landscape in a new and innovative way which has proved to be correct. He lived much of his life in Edinburgh, in a house overlooking Arthur's Seat. Hutton mixed with some of the most eminent scholars of the Scottish Enlightenment, where his novel views caused controversy. But his collection of rocks has mysteriously disappeared, without trace.

ROCK SOLID

James Hutton challenged these ideas on two counts. Firstly, he recognised that there was more than one way of making a rock, which might not always involve the sea. In Frederick Street, Edinburgh, a section of rock was exposed in the side of a drain, and Hutton noticed that a black crystalline type of rock, which he called a whinstone, cut across a sequence of layered rocks. The layered rocks had clearly been laid down in the accepted way in the sea, but the black crystalline rock must have pushed its way in afterwards. It soon became clear that the black crystalline rock had been intruded as a hot, molten liquid which had cooled and solidified. It is in fact an offshoot of lava from one of the volcanoes, which never made it to the surface – a feature which is now known as a **dyke**.

Hutton's second important observation was that Earth processes such as erosion, deposition, earthquakes and vulcanism, are happening all the time around us today, and have happened continuously in the past too. He was the first person to realise that the surface of the Earth, the crust, is active, forever being formed, changed, destroyed and reformed. In Hutton's own words this recycling process had 'no vestige of a beginning – no prospect of an end'. It is this ever-changing crust that makes Earth such a special, and unique, planet in the Solar System. We live on this active crust, and live *from* it – all our needs come from the crust, either by farming its soils, or by quarrying and mining its riches.

Any basic cookery book will tell you that there are three basic ways of making a cake, all of which can be infinitely varied by different flavourings and icings. It's just the same with rocks - three basic recipes, with countless variations for each:

1 Igneous rocks
These are the crystalline rocks, recognised by Hutton as having solidified from a hot melt, either as lava erupted from a volcano, or buried inside the Earth's crust, often underneath volcanoes. The black basalt of Edinburgh's volcanic hills is the most common form of lava, still being disgorged by volcanoes such as Mauna Loa in Hawaii. The crystals in basalt are too small to be seen by the naked eye (this is because the lava cooled very quickly) but they can be seen in their full glory under the microscope.

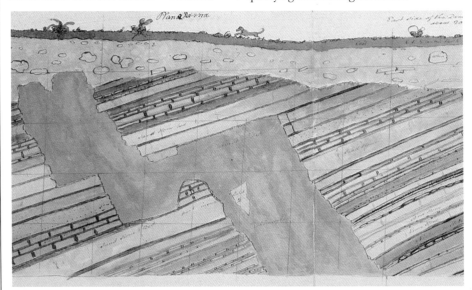

12 A drawing by Hutton's friend, John Clerk, showing the rocks exposed in 'the East side of the drain passing along Frederick Street' in Edinburgh. The little dog at the top gives an idea of scale.

13a Basalt under the microscope.

Quite often, though, lava does not make it to the surface and gets trapped within the crust. Insulated by the surrounding rocks, it cools very slowly and large crystals form, creating some spectacularly beautiful rocks.

13b Gabbro, a coarse-grained version of basalt which has cooled and solidified underground.

2 Sedimentary rocks

Most sedimentary rocks are laid down as sediment under water. They usually occur as layers or beds and frequently contain fossils. They are made from fragments eroded from older rocks (such as sandstone or shale) by the action of wind and water, or from the remains of the shells and skeletons of living organisms (limestone and chalk). So, in a sense, sedimentary rocks can be thought of as second-hand rocks!

It is hard to imagine how a pile of broken shells or sand on the beach can be transformed into hard rock. Indeed, the process isn't fully understood – but the squashing effect of being buried under piles of younger sediment is

14 Sedimentary rock usually occurs in layers (beds) and often contains the remains of living organisms (fossils).

important, and mineral solutions often percolate through buried sediment, leaving behind silica, lime or iron minerals which act as a cement, sticking the grains together.

3 Metamorphic rocks

These are rocks that have been altered by heat, or by a combination of heat and pressure on rocks buried deep inside the Earth's crust. The appearance and texture of the rocks change, and new minerals form, but this all happens in the solid state, without the rock melting.

The rocks immediately next to the dyke in fig 12 will have been baked by the hot molten rock, so have become metamorphic rock. The highlands of Scotland are formed almost entirely of metamorphic rock, squashed and heated

15 Slate is a common variety of metamorphic rock.

during several mountain building episodes in Britain's early geological history.

Two common metamorphic rocks are slate, formed from heated and squashed mudstone or volcanic ash, and marble which is metamorphosed limestone.

Over the immense lengths of geological time, all three types of rock can be transformed into one another in a never-ending story – **The Rock Cycle**.

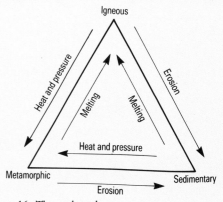

16 The rock cycle.

ROCK SOLID

STAN WOOD – THE MISSING LINK MAN

Stan Wood has a remarkable sixth sense, or a "nose" as he puts it, for finding fossils. He started fossil hunting when he worked as an insurance salesman in his native Edinburgh, and became so successful at finding rare fossils in unusual places (beside bus stops, drystone walls, football fields – he literally left no stone unturned) that he turned professional – collecting fossils and selling them to museums.

Fish

The first discovery that really excited and inspired Stan was a fossil fish he found in oil shale on the shores of the Firth of Forth near Edinburgh. He went on to uncover well over 500 specimens, many of previously unknown species of fish, in other parts of Scotland's Midland Valley. *Stethacanthus*, 'the Bearsden Shark', whose weird appendage carried on its back, like a shoe brush covered in teeth, is still a mystery. It could have been used to catch prey, for defence, or for sexual display – fossils can tell us a remarkable amount of detail about an animal's anatomy and physiology, but reconstructing behaviour is always a matter of educated guesswork.

Missing Links

Fossil land animals also fascinated Stan, and a walk with his dog led to his most important discovery – a discovery which was to rewrite pages of the history of life on Earth. It was 1984, and on the outskirts of Edinburgh, his dog stopped by a drystone wall. Stan immediately recognised that the rock was unusual, finely layered, like the oil shale he was used to, but harder and with fine bands of lime between the paper-thin grey and brown layers. Very soon, Stan had found a fossil which workers at the Royal Scottish Museum in Edinburgh recognised as a very early amphibian. He painstakingly traced the source of stones in the wall back to a tiny quarry which he was given permission to re-open. It turned out to be an Aladdin's cave of rare finds including the most important fossil unearthed this century, the world's

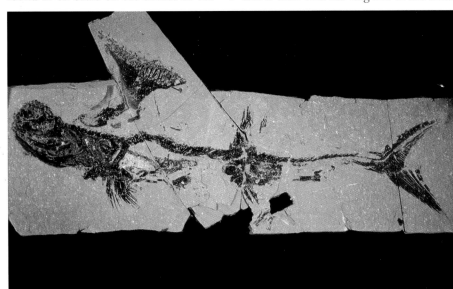

17 The 'Bearsden Shark' fossil from Glasgow.

18 Stan Wood with his dog, Cherry.

oldest reptile, affectionately known as 'Lizzie the Lizard'.

Three hundred and forty million years ago, this quarry had been a lake near the site of a hot spring, fuelled by the heat of the nearby Edinburgh volcanoes. Animals fell into the water – either when they went there to drink, or being chased by predators, or perhaps fleeing from the falling ash and forest fires caused by volcanic eruptions. It would have been a quick death, followed by a rapid burial in mud and lime, and so to fossilisation. The whole range of land animals is preserved, both predators and prey, together with leaves and seeds from the forests in which they prowled: the world's oldest harvestman 'spider'; one of the oldest land-living amphibians

(the ancestor of frogs and salamanders); huge land scorpions up to half a metre long; and the 20 centimetre *Westlothiana lizziae*, 'Lizzie the Lizard'.

There is in fact some doubt as to whether 'Lizzie' is a true reptile, but bones in the skull and ankles are very reptile-like, and almost certainly represent an animal which laid its eggs on dry land and contained the embryo within a membrane, like that inside the shell of a chicken's egg. It is tempting to speculate that 'she' (although the most complete skeleton known may well be male) may be the oldest known member of the group from which not only the dinosaurs evolved, but human beings too, and, last but not least, Stan Wood's little dog.

21 A reconstruction of the site where 'Lizzie' died 340 million years ago.

19 A paperweight made from the finely layered oil shale in which Stan Wood found some of his most important fossils.

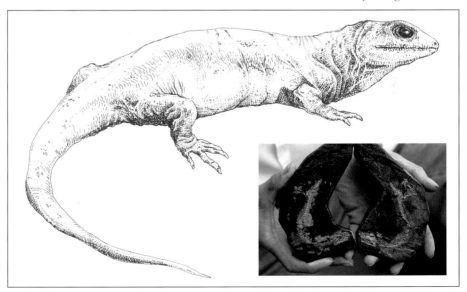

20 'Lizzie the Lizard' – one of the most important fossil finds of the twentieth century.

13

ROCK SOLID

THE SCOURING OF THE ICE

By two million years ago Edinburgh's landscape had changed dramatically – gone was the lush tropical growth of 340 million years ago, and in its place a harsh icy waste. Edinburgh's landscape would have resembled Antarctica's today. Britain was in the grip of the Ice Age.

A huge ice-sheet waxed and waned over a Britain whose outline was beginning to look recognisable. The climate would periodically get warmer and the edge of the ice-sheet would retract northwards. These warmer periods were called **interglacials**, and it was during these times that people came to populate Northern Europe. We are, in fact, in the middle of an interglacial now. The last ice-sheet waned 10 000 years ago, so we can expect another in a few thousand years time. A return to Carboniferous conditions and lush tropical holidays in Edinburgh seems unlikely, but there is great potential for the city as a future ski resort.

There would have been no risk of volcanic hazard, if one had once again time-travelled back to the summit of Arthur's Seat; just the risk of frostbite or losing one's way in a white-out. The Firth of Forth would have been a deep channel, with the odd floating iceberg. The snow-clad Highlands would have been just visible on the horizon, just a little higher than they are today. The ice-sheet moved sluggishly from west to east, scouring the landscape on its way. It was not the ice itself that did the scouring, but a myriad of fragments of rock, gravel

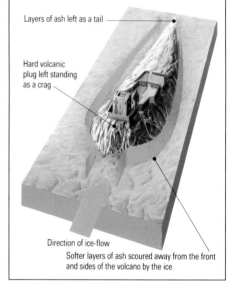

Layers of ash left as a tail

Hard volcanic plug left standing as a crag

Direction of ice-flow

Softer layers of ash scoured away from the front and sides of the volcano by the ice

23 The 'crag and tail' structure of Edinburgh Castle.

22 Two million years ago, Edinburgh would have looked rather like Antarctica does today.

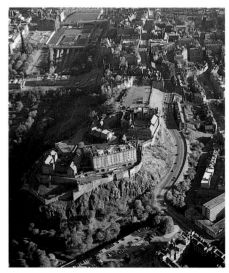

24 The 'crag and tail' today, viewed from the air.

and sand picked up by the ice on its slow journey. Harder, more resistant rocks were left standing, the softer rocks eroded away. It is the the action of the ice that has made the basalt volcanoes of Edinburgh as prominent on the landscape today as they were at their dawn during the Carboniferous.

To the west, north and south, the slopes of the Castle Rock volcano have been swept cleanly away, to leave the bare, much harder rock that had, millions of years previously, congealed in the volcano's neck. To the east, however, in the lee of the ice's path, the gentle slopes of softer volcanic ash remain, making a convenient and imposing drive and entrance to the castle.

In Holyrood Park, on the slopes of the Arthur's Seat volcano, are several *roches moutonnées.* These are low, isolated rocks with a streamlined form – smooth and gentle on the slope facing the upstream of the ice flow, but steep and jagged on the leeward side. As the ice advanced up the rock, it would have sandpapered the surface to a smooth finish, but as it moved over and pulled away from the other side the ice would have taken with it jagged blocks broken from the steep side. The name *roche moutonnée* stems from the fact that, from a distance, they do look remarkably like sheep. Many a geologist has suffered embarrassment when a *roche moutonnée,* so neatly sketched in his field notebook, has risen to its feet and walked off.

What goes up must come down, and what is picked up and transported by an ice-sheet must end up getting deposited somewhere. Some was deposited directly from the ice as it moved along and some was dumped after the ice melted. These ice deposits have dictated the final shape of the lowland landscape – low, smooth, elongate hills, called drumlins, of glacial clay and stones, and irregular hummocky hills of sand and gravel deposited by meltwater streams underneath the ice.

After the ice melted, Scotland was no longer weighed down by the huge amount of ice, and the land started to rise and, indeed, is rising still, at a rate of a few millimetres a year. Southern England, however, was not covered by the ice-sheet, and is now slowly sinking. There are those who would argue that the rocks are trying to make a political point . . .

25 *Roches moutonnées* on the slopes of Arthur's Seat.

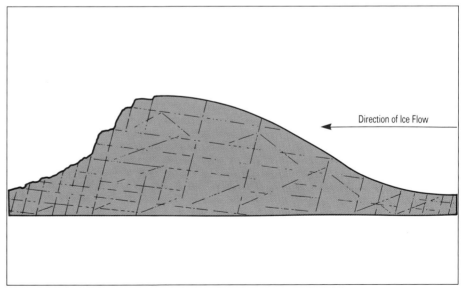

Direction of Ice Flow

26 The formation of *roches moutonnées.*

ROCK SOLID

ENVIRONMENTAL GEOLOGY

James Hutton made Edinburgh the birthplace of British geology, and it is still home to several geological institutions, including a large contingent of the British Geological Survey (BGS). Their research, monitoring, surveying and mapping work is crucial for planning the way we use the patch of crust that we live on.

The map (fig 27) shows the geological factors that will affect future land use in the Stirling area (about 50 km west of Edinburgh). The broad yellow band is soft unconsolidated sediment that was scraped up by the glacier, then deposited in the then wider and deeper estuary of the Firth of Forth. It has long since risen to become dry land.

Dry, but not firm – the wise man would build his house, factory or power station on the purple splodges, which represent the hard rocks of volcanic origin, or possibly on the white area, which represents rocks formed in the Carboniferous coal swamps. The foolish

27 A map compiled by the British Geological Survey showing land use around the town of Stirling, west of Edinburgh.

man, the Bible tells us, built his house on sand. Sand is, in fact, not the worst material for foundations, but it would indeed be a foolish man (or woman) who chose to build on silt or mud in the yellow area of the map without a great deal of care, planning and engineering. A very detailed survey would be required before building a large structure in that area, and it is quite likely that piles would be needed to be sunk deep into the sediment.

The rows of black arrowheads show steep slopes which could be unstable. Many are around the volcanic hills, where the rocks have natural joints and cracks in them, and it is well within the realms of possibility that a sizeable chunk could become dislodged and cause mischief to people or property below. (Fig 28 shows an outcrop of rock with a wall built on top. The joints in the outcrop have been cemented.) Some arrowheads mark the slopes of hills made of material dumped by the glaciers, and there is a line of steep slopes along what was once the cliff-line of the Firth of Forth, when the land lay very much lower.

The green areas on the map show areas of quarrying, waste-tipping, or other places that have been disturbed, or the ground artificially remade, by people. We are a frighteningly powerful geological force ourselves, and judging by the quantity of green on the map, and the timescales involved, we seem to be moving sediment around at about the same rate as the glaciers.

Artificial levées have been built along the banks of the rivers to stop flooding; and on the inside curve of two of the meanders, the low-lying land has been used as a rubbish dump. With time the rubbish will decompose and may form methane – a gas which can be a useful resource, but in the wrong place and poorly managed it can cause horrendous explosions.

Gravel deposited by meltwater streams during the Ice Age is an excellent source of aggregate for concrete, and some of the rocks from the volcanic hills make good road-building materials. So, it's not all doom and gloom, because this map highlights resources as well as hazards to planning.

28 A wall built on volcanic rocks which have been cemented along the joints.

ROCK SOLID

EARTHQUAKE MONITORING: CLOSE TO HOME

The British Geological Survey in Edinburgh includes an earthquake monitoring centre. Their equipment picks up the vibrations from earthquakes all over the world, including Britain. This may come as a surprise, because most of us probably think that Britain is immune from earthquakes, but in living memory two sizeable earthquakes have been felt in central Britain, and smaller tremors are regularly detected by seismographs, even though they are too small to be felt by people.

Earthquakes are caused by the movement of rocks along faults in the crust, or in the mantle which underlies the crust. Most major earthquakes occur

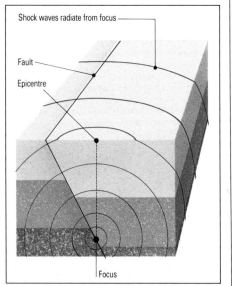

29 The epicentre and focus of an earthquake.

along the present boundaries of crustal plates. Because there are no active plate boundaries running through Britain, we are not shaken by devastating 'quakes. But minor 'quakes still occur along ancient fault lines.

On 2 April 1990, BGS seismographs registered a trace of a 'quake measuring 5.1 on the Richter scale (fig 30). The Richter scale of earthquake magnitude measures the energy released by an earthquake, not the effect it has on people and property at the surface. In fact, two earthquakes of the same magnitude on the Richter scale can have very different effects at the surface.

The distance of the **epicentre** of the 'quake from the monitoring station is calculated from the time that passes between the arrival of the two main peaks on the seismogram. Comparing traces from other seismic monitoring stations pinpointed the epicentre of the 'quake as Bishop's Castle in Shropshire. No one was hurt, although 'quakes of similar magnitude have claimed the lives of many abroad (for example, the 1960 Agadir 'quake in Morocco, which claimed 12 000 lives). This is because the **focus** of the earthquake was very deep (that is the place where the earthquake originated, in the rocks directly below the epicentre), so the effects at the surface on people and property were less pronounced. The effects of a 'quake on people is measured by people: fig 31 shows a map compiled from questionnaire data sent in by the public to BGS.

Earthquakes of this magnitude and intensity are rare in Britain, but sufficiently frequent to require being taken seriously by construction engineers. An earthquake of magnitude 5.3 and of similar intensity (that is, no-one was hurt and damage was slight) occured in Wales in 1983. Again, the focus was very deep.

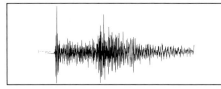

30 A seismogram of the Bishop's Castle 'quake of 1990.

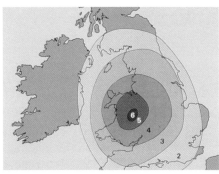

31 The intensities (on the Modified Mercalli scale) felt by people living at various distances around the Bishop's Castle earthquake. The scale runs from 1 to 12. The maximum intensity at Bishop's Castle was 6 — felt by all, some frightened; a few plaster cracks. 5 – felt outdoors, sleepers wakened, liquids disturbed or spilt. 4 – vibration like the passing of heavy trucks. 3 – felt indoors, hanging objects swing. 2 – felt by people at rest. 1 – not felt.

WHAT EARTHQUAKE WAVES TELL US ABOUT THE EARTH'S INTERIOR

Earthquakes are monitored by seismic stations round the globe and in the early days of monitoring, certain kinds of earthquake waves appeared not to be able to travel through the centre of the Earth. Stations on the opposite side of the world seemed to be in a 'shadow zone'. It was realised that these particular earthquake waves could not travel through liquids, and so it was deduced that the outer part of the Earth's core was liquid.

Many people have the idea that the Earth is rather like an orange filled with boiling custard – a thin crust with hot lava bubbling all the way through inside. Occasionally, the crust ruptures and the boiling custard escapes through a volcano. Seismic evidence (as well as evidence from volcanoes and meteorites) has revealed this not to be the case. For the truth see fig 33.

Seismic monitoring has also helped our understanding of volcanoes, because underground reservoirs of molten rock can be detected. It has been found that some of these underground reservoirs, which feed volcanoes at the surface, are in the crust, and some are up to 150 km deep in the mantle. It is these very deep reservoirs that, in the main, have produced the dark basalt rocks like the ones found around Edinburgh.

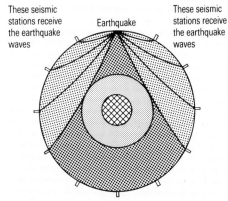

These seismic stations receive the earthquake waves

Earthquake

These seismic stations receive the earthquake waves

Seismic stations in this **shadow zone** receive none of the earthquake waves which can't travel through liquids

32 Some earthquake waves cannot travel through the core of the Earth.

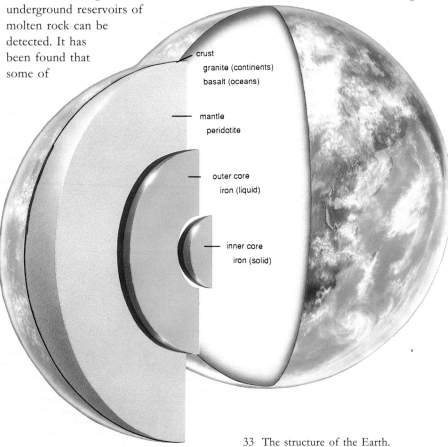

crust
granite (continents)
basalt (oceans)

mantle
peridotite

outer core
iron (liquid)

inner core
iron (solid)

33 The structure of the Earth.

ROCK SOLID

34 Trilobite fossils, 470 million years old.

THE STORY OF LIFE
IN A NORTH – SOUTH DIVIDE

When fossils like those in fig 34 were found in Wales in the seventeenth century, they were thought to be the remains of flat fish. Their shape approximates to a small rounded dab or flounder, and one could imagine them having once been made up of flakes of juicy white flesh. But the segmented nature of the body was soon discovered to be far more akin to woodlice than to anything tasty from the fishmongers – these fossils are, in fact, the hard carapaces moulted and discarded by one of the earliest forms of arthropod, the **trilobites**.

Fossils are organic remains, buried by natural processes and permanently preserved. Usually it is the hard parts, the skeleton or the shell of an animal that are preserved, and only very occasionally are remains of soft parts found in rocks. Sometimes the animal itself is not found, only tracks or burrows it made in the sediment survive in the rock. In order to be preserved, animals and plants have to end up in places where they can be buried before disintegrating or being eaten – countless creatures and plants must have come and gone on this Earth without leaving so much as a trace. So, fossils do not tell the whole story of life – they are more like clues, like the fingerprints, headless bodies and bloodstains of the detective novel which, pieced together, build up a picture of what an organism looked like and how and where it lived.

The trilobites in fig 34 were found in black **shales**, a muddy rock that splits easily. (Shale is a little bit like slate, but slate is harder and splits more evenly. Slate is a metamorphic rock and is hard because it has been subjected to a thorough baking and squashing. Shale is a sedimentary rock, and nearly all fossils are found in sedimentary rocks.) Black shales are often formed in fairly quiet deep water, for example, on the ocean floor. This suggests that Wales was once covered by an ocean and that trilobites were marine organisms that lived in that ocean.

Detailed and painstaking work by palaeontologists has confirmed that about 470 million years ago, the trilobites to which the carapaces in fig 34 belonged, were indeed scuttling around the floor of an ancient ocean, long since disappeared. This ocean, called *Iapetus*, separated the north and south of Britain into two blocks, both lying south of the equator. Wales was roughly where the Falkland Islands are now.

Fossil evidence has shown that *Iapetus* was teeming with life: some, shelly creatures almost indistinguishable from those alive today; others, such as the trilobites, are completely extinct.

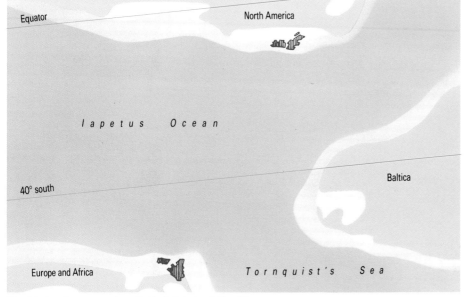

Equator

North America

I a p e t u s   O c e a n

Baltica

40° south

Europe and Africa

T o r n q u i s t ' s   S e a

35   *Iapetus*, the ocean that divided Britain 470 million years ago.

ROCK SOLID

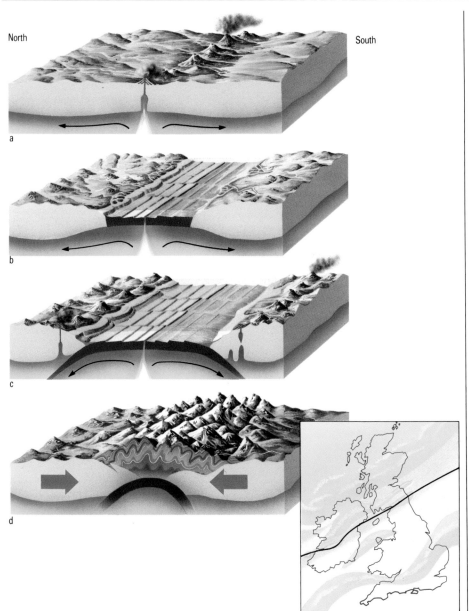

North

South

a

b

c

d

36 The opening and closing of the ancient *Iapetus* Ocean.

THE LIFE AND DEATH OF AN ANCIENT OCEAN

About 700 million years ago, a line of rising convection currents in the mantle caused a long crack to appear from east to west in the Earth's crust. The sections of crust either side of the crack slowly moved apart and new crust, in the form of molten rock (lava) bubbled up in the middle (fig 36a).

The two sections of old continental crust moved further and further apart – very slowly, at about the speed your fingernails grow – riding on the back of the convection currents. Black basalt lava continued to bubble up from the crack, hardening and forming new crust. This, too, rode on the back of the convection currents and moved sideways, while new material constantly bubbled up to take its place.

Gradually, water flowed in over the new basalt crust and an ocean was born. The old blocks of crust remained as continents, one north and one south (fig 36b). The remains of the northern continent are still to be found in Northwest Scotland, home of the most ancient rocks in Britain.

Iapetus probably grew to be about 5000 kilometres wide. But oceans are not stable or permanent features. They grow and then they shrink again. The black basalt crust of the ocean floor took a nose dive and followed the convection currents back down into the Earth causing earthquakes. These areas, where ocean floor is swallowed up, are called **subduction zones** (fig 36c).

As the crust dives down into the Earth, it heats up. Some of it melts because of the friction, the molten rock coming back to the surface to form volcanoes. This is currently happening on both the east and west sides of the Pacific Ocean, along the west coast of South America and in the Philippines and Japan. This means that the Pacific Ocean is slowly closing up.

About 500 million years ago a subduction zone developed running west to east through what is now North Wales and part of the Lake District, along the southern side of *Iapetus*. Another subduction zone had developed on the northern side of *Iapetus*, and the volcanic rocks formed along it can be found in the Southern Uplands of Scotland. Slowly, the great *Iapetus* ocean began to shrink, and the two ancient continents moved closer together.

Eventually, the continents collided – not in a loud and sudden cataclysm, but slowly, and forcefully. Sediments which had accumulated in the deep troughs above the subduction zones were pushed together, folded and scrunched up (fig 36d).

The two ancient continents, originally thousands of miles apart were now welded together. The join, rather amazingly, runs approximately under the line of Hadrian's Wall – but the Romans who built it were doubtless unaware that Scotland was once part of a different continent from England and Wales (fig 36e).

The Atlantic Ocean is now at about the same stage *Iapetus* was before it started to shrink. We are moving away from North America (geologically at least) at a rate of a couple of inches a year, and one fine day, it will start to close. It has been suggested that the devastating Lisbon earthquake of 1755 and the Agadir 'quake of 1960 were

The shrinking and closure of *Iapetus* formed rocks that have been vital to the economy of Mid- and North Wales.

Slate was created when fine-grained rocks formed from muddy sediment on the sea floor were heated and squashed (**metamorphosed**) as the two continents collided.

Slate is waterproof and can be split into thin sheets along the **cleavage**, which is formed when clay minerals in the mudstone recrystallise into microscopic flakes of mica that grow in layers at right angles to the

precursors to the development of a subduction zone, but there is no firm evidence for this. Nonetheless, in another couple of hundred million years, Europe and Africa may well find themselves welded onto the Americas.

It is, however, unlikely any humans will be there to watch it happen or to reap any political benefits.

direction of the pressure. You can reproduce a similar sort of effect by pushing a random pile of matchsticks – they will align themselves at right angles to the pressure.

The volcanic rocks formed above the subduction zone are a very important source of roadstone. The top surface of a road needs to be made from chippings of a rock that is tough, won't absorb water and, most importantly, is skid-resistant – a rock that won't polish easily under the constant rubbing of tyres.

37 Splitting Welsh slate.

38 A roadstone quarry near Builth Wells.

ROCK SOLID

THE DAWN OF LIFE

No-one knows for certain how life began, but there is evidence to suggest that life may have had its beginnings quite early on in the Earth's 4600-million-year history. Some of the oldest rocks in existence (about 3800 million years old – the very oldest rock so far found is 3960 million years old) contain a type of carbon known to be associated with organic activity. Carbon is the stuff that life is made of – animals, plants and bacteria. We are all basically a complex mixture of carbon chemistry. Life also needs water for sustenance, and a few other elements, such as nitrogen and phosphorous, in a fairly narrow temperature range (0°C to 100°C, the range at which water is liquid). The Earth provides all this, because it is neither too far from the Sun nor too near, and large enough to have a gravitational field strong enough to hold on to its oceans and all-important atmospheric blanket of gases.

Tantalisingly, some meteorites have been found which contain simple carbon chemicals of the sort that are found in living organisms – so perhaps life came to Earth from space. We shall probably never know the exact process, but somehow carbon chemicals started to form in the primeval oceans, and made the step towards becoming cells. This was a gigantic step – cells are the building blocks of life and home to DNA, the blueprint of life.

That life started in the sea seems indisputable, and fossils of single cells, such as bacteria, have been found in very ancient sedimentary rocks, well over 3000 million years old. Single-celled organisms appear to be all that lived on the Earth for about 2500 million years, until about 650 million years ago.

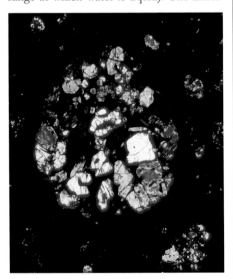

39 A slice of carbonaceous chondrite meteorite. The carbon is black.

40 One of Britain's oldest fossils of a multicelled animal – a jellyfish from the Ediacara fauna.

Some of the oldest fossils of recognisable animals in the world, the Ediacara fauna, have been found just south of Carmarthen in Wales. They are the marks made by delicate sea-pens and jellyfish-like animals as they sank on to the fine sediment of the early *Iapetus* ocean. As the sediment hardened into rock, these 'death-masks' were preserved. But it is doubtful whether all these early creatures are ancestors of later life; some at least are probably the remains of unsuccessful attempts that did not survive.

The Explosion into Life

Five hundred and seventy million years ago, life as we know it really seemed to take off. From this time on an abundance of fossils appears, a marked change from older rocks in which fossils are extremely rare. Animals at this time had developed the sort of shells and skeletons that were hard enough to be preserved. There may have been a lengthy gradual evolution of soft-bodied animals before then, which by their nature were not preserved. But most scientists believe that, for some reason yet unknown, there was an abrupt explosion of life, and the seas were suddenly teeming with a huge new range of different creatures.

All this happened in an age called the **Cambrian**, named after the Cambrian mountains of Wales, where these abundant fossils were first found. The range of body plans of animals in the Cambrian was as diverse as is found in the oceans today. Some sponges and shelled animals are much like present living forms, and may well be their ancestors; others are stranger creatures, now extinct, with little similarity to modern animals.

41 A graptolite. These fossils are common in Wales, and were once colonies of living creatures of a type long-since extinct.

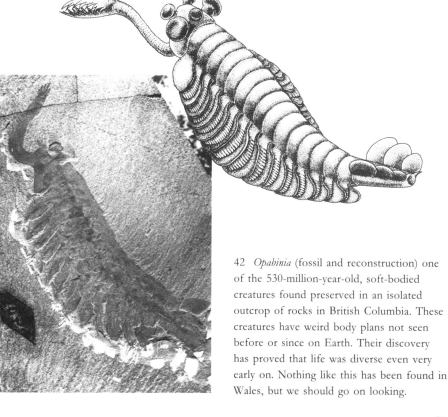

42 *Opabinia* (fossil and reconstruction) one of the 530-million-year-old, soft-bodied creatures found preserved in an isolated outcrop of rocks in British Columbia. These creatures have weird body plans not seen before or since on Earth. Their discovery has proved that life was diverse even very early on. Nothing like this has been found in Wales, but we should go on looking.

ROCK SOLID

THE EVOLUTION DIARY

Darwin's seminal work *The Origin of Species* was published in 1859. It established the idea that animals and plants evolved from their ancestors, and that if an organism grows with a new characteristic that gives it an advantage in its environment, it will stand more chance of surviving and producing offspring, which in turn may inherit that advantage.

Darwin also recognised that species could change if they became isolated from others – on an island for instance. The importance of geographical separation is evident from the fossil record – for example, trilobites found in Northwest Scotland are of the same types as those found in North America, but very different from those of the same age found in Wales. This is because Scotland was once joined to North America, yet was separated by an ocean from England and Wales, so the trilobites that inhabited the shelf seas bordering continents were isolated from one another and had evolved into different species.

One aspect of Darwin's evolution theory has proved troublesome: the idea that organisms evolved slowly and gradually, rather than in sudden leaps and bounds. Until recently, there was almost no fossil evidence for this – fossils found in one bed seemed markedly different from others in the beds beneath. It was assumed that the sudden changes found in a sequence of rocks were because not all the layers had fossils in them; or that many hundreds or thousands of years might have passed without a bed of rock being laid down; or that some beds may be missing because they were eroded away before the next lot was deposited. In other words, the fossil record is incomplete, rather like an old diary in which some the pages have not been written or while others have been torn out altogether.

But in the 1970s, some American palaeontologists pronounced the theory that 'missing links aren't there because they weren't preserved' as unscientific, and a limp explanation. They challenged Darwin, and proposed an alternative theory: that a species carries on much the same for a while, and then suddenly changes into a different species – maybe triggered by a sudden environmental change.

Then, in the mid-1980s a British geologist, working in streams just outside the Welsh spa town of Builth Wells, found evidence in the rocks that evolution can, and has, happened gradually. Peter Sheldon had great trouble in allocating the correct Latin names to the hundreds of different trilobites he was finding in the black shale exposed in the streams – some had too many ribs in the tail, others too few. Some lacked the knobbly bits in the head which he expected to find.

43 Dr Peter Sheldon in one of the streams which revealed to him some secrets of evolution.

44 The word trilobite means three lobes – the body is divided laterally into three.

Many people would have been tempted to write down the name that was the nearest fit, then nip down to the *Leek and Dragon* for a pint. Not Peter Sheldon; he recorded every minute detail of the knobbly bits and numbers of tail-ribs, alongside the exact rock horizon that each trilobite came from. After a computer analysis of 15 000 specimens, he found that there were gradual changes in the numbers of tail-ribs and knobbly bits in many lines of trilobites.

Sheldon had proved Darwin right, at least in this case, where pages of the Earth's 'rock diary' are relatively complete. He showed that it is often the way in which people collect and ascribe names to their fossils that gives an artificial impression of sudden change and obscures evidence of gradual change.

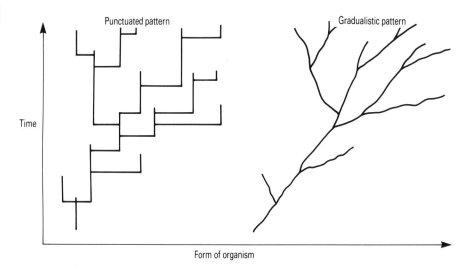

45 Alternative shapes to the tree of life. On the left, a punctuated pattern in which all the changes happen suddenly and quickly; on the right, change is continuous, one form gradually changing into another.

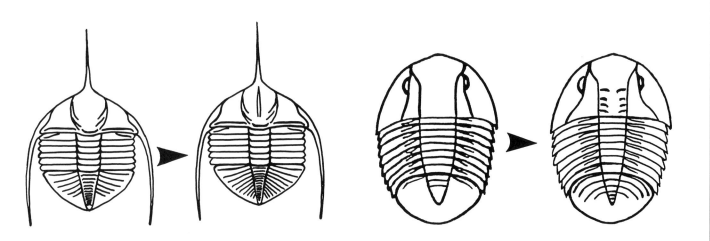

46 Spot the difference: changes in two trilobite lineages from central Wales occurring over about two million years.

ROCK SOLID

47 *Cooksonia pertonii* — the oldest land plant found in Britain, and one of the oldest in the world.

48 *Gosslingia breconensis* a very early plant that was common in Wales 410 million years ago.

STEPPING ON TO THE LAND

Many people assume that the leap of life on to land from water was sudden and dramatic. This is unlikely. It is now thought that green smears of algae and bacteria may have covered some surfaces not all that long (in geological timescales of millions of years!) after life began. All life needs water, in seas or rivers or on dry land — so these primitive green microbes would have needed high humidity and water-logged soils to keep moist. It is likely that they would have been confined to sand, rocks or mud by the water's edge. For life to venture far from a regular supply of water, a method of retaining water and preventing water loss was needed. Most modern land plants are covered with waterproof waxes, have a root system and contain water- and food-carrying tubes called a **vascular system**.

Some of the oldest land plants in the world have been found just north of the Brecon Beacons by Dianne Edwards, a botanist working at the University of Wales in Cardiff. In an unassuming little quarry containing rocks laid down 420 million years ago, she found even more unassuming black streaks in the rock (fig 47). Each one is a simple branching stem just a few centimetres long with

49 A reconstruction of the earliest land plants; just as green as grass is today.

each branch ending in a sporangium – a sac containing spores. The stems have a waterproof coating pierced with small holes (stomata) to allow gases (carbon dioxide, oxygen and water vapour) in and out. The sporangia would burst and the spores disperse in the wind. This wind dispersal system was convincing evidence that these little plants, *Cooksonia pertonii*, grew on dry land. The fact that it had stomata suggests very strongly that it was a true land-living, vascular plant – the earliest in Britain, and one of the first in the world.

In slightly younger rocks, 410 million years old, Dianne has found the remains of plants in which every minute detail of cell structure and a water-carrying vascular system is preserved. *Gosslingia breconensis* (figs 48-50) probably grew in large green swathes on the flood plain of a braided river system, looking much like low reeds or tall grass would today. Life crept not from the sea on to the land, but from fresh water. The sinuous courses of rivers intertwined the worlds of water and land – there was plenty of land near water and plenty of opportunity for plants to take hold.

It would be a mistake, however, to assume that plants got there first and were followed sometime later by animals. In all likelihood, both made the move at roughly the same time. The early algal smears may well have been foraged by animals – evidence for this comes from worm burrows found in rocks of about 450 million years old. The animals breathed the oxygen produced by plants,

and exhaled carbon dioxide, which in turn was taken in by plants to produce sugars and oxygen again.

Very recently, in the Welsh borders, fragments of fossil centipedes and spider-like animals have been found in rocks of the same age as the early vascular plants found elsewhere. They probably fed on the litter of dead plants or on other smaller invertebrate animals. These early litter feeders, decaying plants and their root systems, the fungi associated with them – plus air, water and fragments of rock – would have combined to produce a substance fundamental to the evolution of higher life on the land – soil.

51 Knee joint of a centipede (x60), found in rocks 414 million years old. Fossils such as this suggest that plants and animals colonised the land at about the same time.

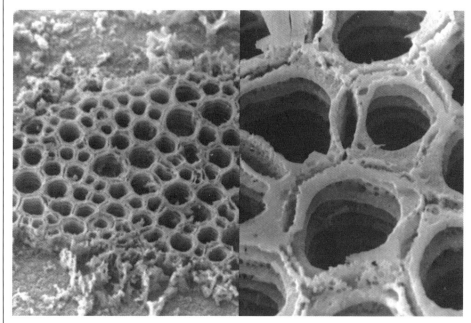

50 Every minute detail of the cell structure of these early plants is preserved in the fossils, as these electron microscope pictures of *Gosslingia* show.

ROCK SOLID

52 Limestone countryside viewed from the National Stone Centre in Wirksworth, Derbyshire.

BRITAIN'S HEARTLAND

The Pennines, stretching from Derbyshire in the south to Northumberland in the north, are the backbone of England in terms of landscape, culture and industry. The hills and dales of Peak District National Park make it one of the most popular inland tourist areas in the country; the Pennines have formed a cultural and political divide throughout the centuries, from the Wars of the Roses to contemporary conversations about cricket. But above all, the Pennines are an economic backbone: it was the rocks in and around the Pennines which fuelled the Industrial Revolution and made Britain's economy and place in world politics during the past 200 years anything but spineless.

All the rocks of the Pennines can be put to good use. Each layer in the sequence of sedimentary rock laid down between 360 and 290 million years ago, in the **Carboniferous** period, has played a part in human history. At the start of the Carboniferous, Britain was at the equator. Scotland was a tropical paradise, while most of England and Wales was submerged under a warm, clear sea.

The creatures that made up that warm, tropical sea had shells and skeletons made of calcium carbonate (just like modern sea-shells). When they died the shells and skeletons sank to the bottom of the sea, to join an accumulating pile of like material. After thousands of years of burial and millions of years of being squashed under several hundreds of metres, not to say a few kilometres, of sediment, the shells and skeletons were transformed into hard limestone.

Sometimes the creatures are still clearly visible, preserved as fossils in the limestone. Fig 53 shows a kind of limestone, from nearby Nottinghamshire, that is made up entirely of fossil corals. Very often, though, the shells and skeletons break up into small fragments as they are moved about the sea bed and knocked against one another by currents. In much of Derbyshire, algae formed a fine calcium carbonate mud which has become a uniform pale grey rock, made up of particles so fine that its organic origins are only recognisable under the microscope. The distinctive cold, pale grey of this limestone is due to its purity;

limestones of other hues contain impurities (the gleaming spires of Oxford, for example, carved out of Cotswold limestone, owe their warm, honey-gold hues to small amounts of iron oxide). The purity of Derbyshire limestone makes it a particulary useful material as an industrial chemical – a chemical that finds its way into our lives in a wide variety of forms. We meet limestone every morning in the bathroom – finely powdered limestone is

the main constituent of toothpaste.

Limestone scenery is distinctive and beautiful. It houses important habitats for plants and animals alike. Human beings and limestone have always gone hand in hand, so important archaeological sites are often to be found in limestone country. The balance between our need to quarry limestone and these environmental factors is obviously a delicate one, and requires careful management.

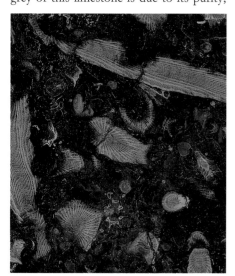

53 Polished Frosterly limestone, full of fossil corals.

54 Quarrying limestone at Tunstead Quarry, one of the largest in Europe.

ROCK SOLID

LIMESTONE – THE CORNERSTONE OF
INDUSTRIALISED SOCIETY

Construction

Powdered limestone is the main
constituent of cement; crushed
limestone chippings are used as
aggregate for the lower courses of
roads. The primary ingredients for
glass are limestone, salt and sand,
while cut blocks can be used for
building.

55

Extraction of metals

Limestone has the property of
lowering the temperature at which
metal can be extracted from an ore. It
is used in the iron and steel industry,
and in the smelting of copper, nickel,
chrome, gold and silver.

Chemicals for industrial manufacture

It would probably be easier to list

56

processes that do not use limestone.
Those that do, and rely on it heavily,
include: plastics, paint, paper, rubber,
car bumpers, dyes, soap, solvents, anti-
freeze . . . you name it, limestone can
do it!

Agriculture and food production

Many plants, particularly cereals, need
an alkali soil, and so fields are limed in
the spring. It is also used to make

57

fertilisers and pesticides. In fish
farming, limestone is added to the
water, and to oyster beds where it
deters starfish from pinching the
oysters. Hens' eggs are made of
calcium carbonate, so limestone is
often added to poultry feed to
encourage strong shells. Limestone is
also used in sugar refining. Without
it, many a delicious pudding, like
creme caramel, would bite the dust.

An environmental 'indigestion tablet'

Limestone has a vital role to play in
cleaning up the environment. As an
alkali, it can be used for neutralising
lake water that has become acidic
through acid rain. Used in power
station chimneys it prevents acid
emissions escaping into the
atmosphere. It can also be used to
remove metal pollutants from streams
and rivers.

58

CAVES, CAVERNS AND CRYSTALS

Derbyshire's limestone is full of holes, caves and caverns – some natural, some man-made and some a bit of both. Although the limestone was laid down in horizontal layers, it has since been bent by Earth movements into an inverted saucer shape. As it bent, it cracked, and it is these cracks, or joints, that formed the starting point for Derbyshire's sub-surface 'cave-scape'.

Soon after the cracks appeared, fluids containing dissolved minerals percolated through them. The minerals were deposited on the walls of the cracks as crystals of silver-grey lead ore (**galena**); creamy barium sulphate (**barytes**); and calcium fluoride (**fluorspar**) in a range of colours from emerald green to deep purple. The fluorspar sometimes occurs as large, spectacular perfect cubes, but more often in Derbyshire as a mush of crystals deposited in layers of varying hues of purple and yellow – this is the famous Derbyshire **Blue John** (fig 61).

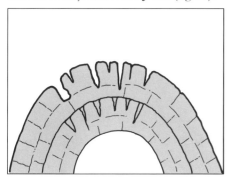

59 Limestone is brittle, so when it is bent into a fold, the top surface cracks.

60a Fluorspar (calcium fluoride).

60b Galena (lead sulphide).

60c Barytes (barium sulphate).

61 A vase carved out of Blue John.

ROCK SOLID

62 Ingots of lead dating back to Roman times have been found in Derbyshire.

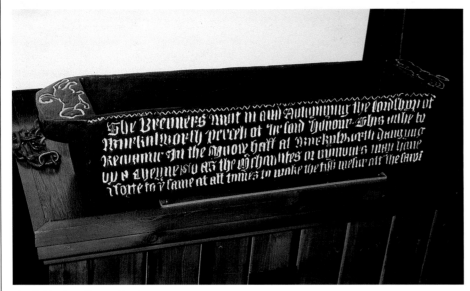

63 A 'freeing dish' of lead ore.

Lead mining was big business in Derbyshire during the Roman occupation of Britain. The Romans used lead for coffins and water pipes, and we still use the word plumber, from the Latin *plumbum* for lead. The industry seemed to disappear into a mega-recession during the Dark Ages, but blossomed again with vigour during the Middle Ages.

The medieval mining customs of Derbyshire are still observed, and the industry is still overseen by *Barmoot*

64 Visitors underground at Temple Mine, Matlock Bath.

Courts, run by a Barmaster and a jury of twelve miners. The barmoot's main function is to oversee mining rights under the Derbyshire Lead Mining Act of 1852, which is still on the statute books, and they still have control over who owns rights to particular mineral lodes. Very little real business is done now, but everyone enjoys a good lunch.

Since the year 1288, anyone discovering a vein and wanting to mine it, has had to make an application to the Barmaster to register the new vein in his book – a process known as *freeing*. The miner was required to pay the barmaster one *freeing dish* of ore (containing about 30 kilograms). From then on the lode had to be continually worked, and various royalties and tithes paid. Stealing someone else's lead ore could result in severe penalties in these earlier times, such as being pinned down by the hand and left to face the choice of death by slow starvation or having to cut off one's own hand to get free.

The verb 'to nick' originates from Derbyshire – if someone wanted to claim a mine not being worked by its owner, the Barmaster would cut a 'nick' in the barrel of the windlass at the top of the mine. If after three such nicks were cut, the original owner had not started to rework the mine, the new claimant was allowed to take possession.

Nearly all the lead and ornamental Blue John has now been worked out from Derbyshire. Some of the old workings left behind in the limestone are now tourist attractions such as the Blue John Caves, Speedwell Cavern and Temple Mine (now part of the Peak District Mining Museum at Matlock Bath). But, tucked away in the hills behind Stony Middleton, attention is turned towards the non-ornamental form of fluorspar – about 80 000 tonnes are still extracted each year. We meet fluorspar every day, in the form of fluoride, mixed with ground up limestone to make toothpaste. Some readers may have been invited, or have issued invitations, to see etchings, printed from glass etched with hydrofluoric acid, also derived from fluorspar. But these are comparatively minor uses (only 0.5 per cent goes into toothpaste). Fluorine is widely used in the manufacturing industry, for example, in non-stick pans, metal extraction, dyes and insecticides. It is used to make refrigerants, both chlorofluorocarbons (CFCs) and other, newer, products designed to be less damaging to the ozone layer.

65 An etching of Matlock High Tor from Chantrey's *Peak Scenery,* first published in 1889.

ROCK SOLID

66 Stalactites growing from the roof of Treak Cliff Cavern, Castleton, Derbyshire.

NATURE'S HIDDEN SCULPTURE

Limestone is soluble in water. So as rainwater penetrates the cracks it widens them by dissolving the limestone. Underground streams and rivers form, hollowing out cave systems. Eventually the dissolved limestone reprecipitates, just like the scale in a kettle, but considerably prettier. The stalactites in fig 66, from Treak Cliff Cavern, Castleton, probably took hundreds or thousands of years to form – but in geological terms, that's fast. (Remember the difference? Stala**g**mites form on the **g**round and *might* grow to reach the top of the cave; stala**c**tites form on the **c**eiling and have to hold on *tight*.)

MILLSTONE GRIT

All good things come to an end. Around 320 million years ago, sand and mud started to wash into Derbyshire's Carboniferous coral sea. A river, flowing from the north was disgorging its load into the sea, forming a delta. The coarse sands of this delta have hardened to become a rock formation called the **Millstone Grit** – so called because it was used to make millstones. The sand grains are made of quartz which is hard, and after the sand was buried and squashed, more quartz formed around the grains, sticking them together. The result is a very tough even-textured rock, excellent as a grindstone.

Millstone Grit was used during the middle Bronze Age (about 1350 B.C.) to make the *Nine Ladies* stone circle on Stanton Moor. Their name derives from the legend that the nine were wicked women who danced on the Sabbath when they should have been in Church and were turned to stone as a punishment. Credance was added to this tale by the fact that their fiddler was also preserved as a single standing stone nearby.

The Millstone Grit was used to build just about every Victorian villa in Matlock and, at the other end of the social scale, Chatsworth House, which is crammed with treasures of Blue John, and whose water gardens are fed by channels lined with lead – truly a monument to the treasures of the Peak District. Millstone Grit is still quarried as a building stone today. The initial cost of building with natural stone is often expensive, but as the *Nine Ladies* have proved, a tough stone can survive the wind and weather for at least 3000 years.

69 Old millstones.

67 Cutting blocks of Millstone Grit for use as a building stone.

68 The *Nine Ladies* stone circle, Stanton Moor.

ROCK SOLID

COAL AND SUNSHINE

Over time, the Millstone Grit delta grew bigger and bigger. Britain was still near the equator, so a lush tropical forest colonised the top of the delta. Trees died, fell over and partially decomposed to form peat. More trees grew on top, and over hundreds of years, thick layers of peat built up. Because the whole area was subsiding under the weight of accumulating sediment, the sea would inundate the delta from time to time, and cover the peat with mud and sand. This sediment would build up, eventually peeping up above the surface of the water, where it would once again be colonised by trees. Subsidence would eventually allow the sea to wash over the delta and the whole process started over again. After millions of years of burial and squashing, the peat became coal, sandwiched between layers of shale and sandstone. A one-metre thick seam of coal would have started out as eight metres of peat, which gives some idea of the intensity of squashing.

The tropical rainforest origin of coal reveals itself in the beautiful fossil ferns found in coal pits (see inside back cover). Preserved in intricate detail, it is easy to imagine them once green and using the Sun's energy to combine carbon dioxide and water into sugars, which fed the plant's growth. That energy is now stored in the coal, and is released again when the coal is burnt, re-releasing carbon dioxide and water into the atmosphere.

There is now a move away from the deep mining of coal in favour of **opencast** extraction. This, in spite of what the papers tell us, is due to geological pressures as much, if not more, than political ones. It is not economic to mine very narrow seams by deep mining, but seams only a few inches thick can be removed by opencast techniques. Many a deep mine has closed when a fault displacing the coal seam was encountered – this again is not so much of a problem with opencast. Very often old, abandoned deep workings are encountered in opencast operations, and coal that could not be removed then (for example, when pillars of coal were left to stop the roof caving in) can be dug out now. Opencasting has considerable benefits for the miner in terms of safety and better working conditions. The only downside to opencast workings is that they can be a blot on the landscape. The environmental aspects – eventual re-landscaping, noise and dust – must be carefully managed, even though it can be costly.

Coal itself makes up only 3 to 4 per

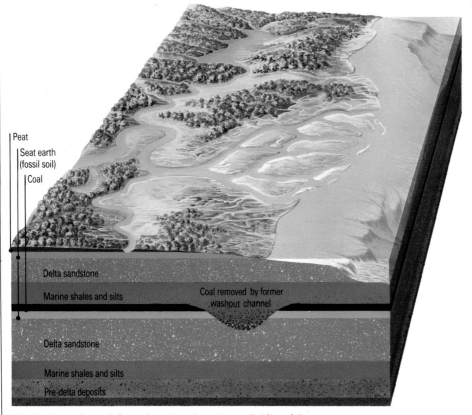

Peat

Seat earth (fossil soil)

Coal

Delta sandstone

Marine shales and silts

Coal removed by former washout channel

Delta sandstone

Marine shales and silts

Pre-delta deposits

70 Coal was formed from decaying plants in a subsiding delta.

cent of the coal measures, so a considerable amount of overburden (the rock on top) has to be removed by diggers and dragline excavators. The last traces of rock are removed by sweeping with a broom, to reveal a black expanse of coal-seam which looks for all the world like a tarmac car park. Overburden rocks are not wasted – all can be put to good use: the sandstone can be used as 'York Stone' paving, or for landscaping the area once the coal has been extracted. Shales and mudstones yield good clay for brickmaking. Underneath every coal seam is a layer of fossil soil, **seat earth**, in which the trees once grew - this is used as a clay for fire-bricks and, in Derbyshire, is used for the famous Denby pottery.

Some of the coal measures in Derbyshire include layers of ironstone. Using the coal, and limestone from nearby, this ironstone was once smelted to forge the iron that built the machines that drove the mills of the industrial revolution. St Pancras station and Vauxhall bridge in London are made of iron from Derbyshire. The ironstone is no longer worked, but its heritage is with us, not least with the 'sunshine miners' who continue to work the opencast pits of the Pennines.

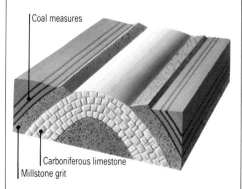

72 A simplified structure of Derbyshire.

71 Opencast miners are nicknamed 'sunshine miners'.

73 The last traces of overburden rock are swept away by hand.

ROCK SOLID

ENGLAND'S ROCKY EXTREMITY

The Cornish landscape is typified by tall, rugged sea-cliffs, and by the ruins of engine houses that contained the huge beam engines which once drove the machinery of Cornwall's mining industry. Both the cliffs and engine houses are made of granite – the latter hewn from quarries in great blocks and laid together with a worthy craftsmanship as solid, robust and pleasing to the eye as any Romanesque fortress. Granite owes its ruggedness and dogged resistance to wind and weather to its **igneous** origins.

Granite is a much paler rock than basalt – this is because it is chemically different, being made of rocks from the crust which have melted and then solidified again, rather than a melt from deep down in the mantle. This solidification took place deep underground, very slowly, which permitted large crystals to form. Sparkling **mica** flakes (which may be either silver or brown) glitter amongst creamy-white **feldspars** and clear or milky-white **quartz** crystals. The crystals are hard and their edges are interlocked, so the whole rock is remarkably tough.

75 Cornish granite.

74 Derelict engine houses at Trewavas Head, South Cornwall.

THE VARISCAN OROGENY:
THE MAKING OF CORNWALL

Four hundred million years ago, when the rest of Britain was dry land, Cornwall and Devon were submerged beneath a deep ocean basin. Thick layers of sediments accumulated, so the floor of the basin sagged downwards. This ocean stretched down to the south of France and beyond, and between the Central Massif and Brittany a **subduction zone** developed (fig 76a).

France was effectively pushing its way into Britain, causing the rocks in the bottom of the ancient ocean trench to heat up, though at first not to the point of melting. This pushing and heating **metamorphosed** the sediments laid down in the ocean basin into greeny-grey slates known locally in Devon and Cornwall as killas. This pushing and shoving caused a concertina effect. Huge wedges of rock were folded, faulted, and piled in great wedges on top of one another. The Lizard peninsula is a slice of ancient ocean floor that was pushed up onto the continent – its **serpentine** rock is quite different from anything else on the Cornish peninsula.

Eventually rocks buried deep inside the squashed pile started to melt (fig 76b). A vast elongated body of molten magma (a red-hot 759°C)

forced its way upwards into the rocks above.

Blobs of magma rose from the main body of molten rock, and cooled to form roughly cylindrical masses of granite which, millions of years later, were exposed to form Dartmoor, the moors of Cornwall and the Scilly Isles. This all happened around 300 million years ago and was the last time that an active plate margin passed through

Britain, although not quite the last time that we have felt the effects of an **orogeny** or mountain-building episode. When they were first formed, the Cornish mountains may have been as high as the Himalayas. Three million years of weathering and erosion of the overlying rocks exposed the granite roots and reduced them to the mellow rounded hills so beloved of dairy cattle and tourists alike.

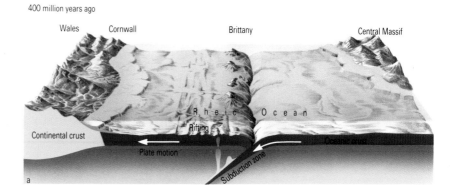

400 million years ago

Wales Cornwall Brittany Central Massif

Rheic Ocean

Rifting

Continental crust

Plate motion

Oceanic crust

Subduction zone

a

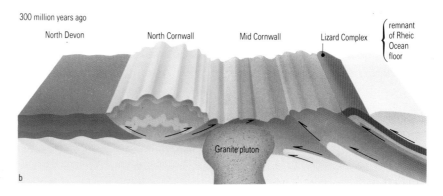

300 million years ago

North Devon North Cornwall Mid Cornwall Lizard Complex

remnant of Rheic Ocean floor

Granite pluton

b

76

ROCK SOLID

A pebble hunt on any Cornish holiday beach can reveal clues to the region's violent past. Pebble (a) is a fold in 'killas' slate; (b) is a fault in which the layers of rock overlap because of the compressive pushing; (c) is a pebble of spotted 'killas' slate that has come from near a contact with granite. The heat of the cooling granite caused spots of the mineral **chlorite** to grow – a sort of geological heat rash! (d) is a chunk of rock from The Lizard – a piece of ocean floor that has been heaved upwards, squashed and metamorphosed into serpentine, which is a common sight in every Cornish gift shop carved into anything from ashtrays to aardvarks.

MINERALISATION

An inter-reaction between the cooling mass of Cornish granite and the surrounding slate around 300 million years ago formed some of the most beautiful minerals and crystals to be found in the world. There was a great deal of water about during the upheaval of the Variscan orogeny, both within the granite and in the surrounding rocks. As the granite was hot, and the surrounding slate cooler, convection currents were created, moving upwards at the margins of the hot granite and downwards in the cooler slate. As it circulated, the water leached metals from the rocks. As it solidified and cooled, the granite shrank and cracked; the slate had its own pattern of cracks in the form of its cleavage and joints formed by the stresses of the pushing movement. Minerals started to crystallise out in these various cracks from the circulating solutions.

There tends to be a zoned pattern of minerals – tin and tungsten closest to the granite, then copper and arsenic, lead and zinc, and finally iron. It is likely that the tin, tungsten, copper and arsenic came from the granite, and that the lead, zinc and iron were leached from the slates. Most workable minerals occur in elongated sheets or **lodes**.

This type of concentration of metals and deposition of minerals by convecting solutions is called **hydrothermal mineralisation**. The quantity and quality of minerals produced this way in Cornwall gave rise to one of the most

a

b

c

d

77

prolific and long-lived mining industries the civilised world has seen. For two thousand years Cornwall's wealth has had a marked effect on Britain's stakes in global economics. But hydrothermal mineralisation has not only fuelled industry and wealth: from out of the darkness and dirt of the old mines of the past few centuries have come some of the most exquisitely beautiful natural objects imaginable. An infinite variety of crystals from bold sparkling prisms to delicate needles, some clear and pure water-white, others all colours of the rainbow; all the lustres from milky-pearl to gold. A visit to one of the many famous Cornish collections is as rich and rewarding an experience as a visit to any art gallery.

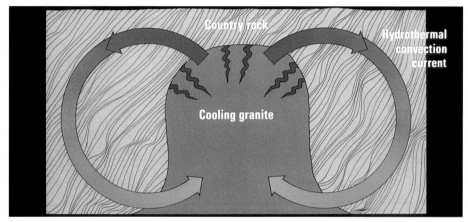

78 Cross-section of the Carnmenellis granite showing the mineralising convection currents.

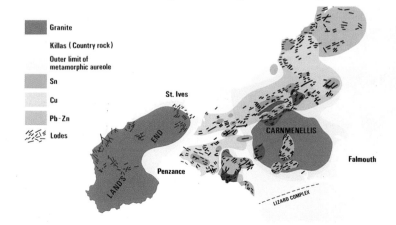

79 Mineralisation around Cornish granites.

80 Turquoise: copper aluminium phosphate.

81 Chalcopyrite: copper iron sulphide.

ROCK SOLID

CORNWALL'S MINING HISTORY

Cornwall is the home and training ground of hard-rock mining. The saying goes that if you visit any metal mine anywhere in the world you will find a Cornishman – 'cousin Jack' – working there. The county owes its particular richness to the presence of tin, which is a comparatively rare metal in the Earth's crust, but valued throughout the ages as a very useful one. Some have claimed that the Phoenicians once sailed to Cornish shores to trade tin, but there is no archaeological evidence for this. Tin may well have been extracted as early as 3000 BC, but a primitive though prosperous industry would have been established by the Bronze Age, around 2000 BC, with a thriving export trade in

82 Cassiterite, the ore of tin in crystalline form.

full swing by 1500 B.C. Bronze, the metal that gave its name to a world-wide phase of human history, is an alloy of copper and tin, and would have been made simply by heating lumps of tin ore (cassiterite), found in stream beds, with charcoal and native copper, in a fire made fierce with bellows.

The Romans developed pewter (an alloy of tin and lead) and also used pure tin. Again, it was likely to have been alluvial tin, washed down into rivers. The heavy black cassiterite pebbles would concentrate where the current was sluggish. True mining of the ore – hewing away at the hard rock – did not become widespread until well into the Middle Ages, as likely as not after lead mining was established in Derbyshire.

From late Medieval times onwards the tin industry expanded and by early Victorian times nearly half the world's tin was mined here. Copper, too, was important, about half of world needs were met by Cornish mines in the early nineteenth century.

83 Cassiterite pebbles panned from a stream.

At first miners would follow an ore lode outcropping at the surface down into the rock. But these shallow 'easy' deposits were soon worked out and shafts were sunk, down which the miners would climb on wooden ladders. They would drive tunnels at intervals into the lode, which would be worked in such a way as to leave as little ore as possible behind. As mines got deeper and deeper, there was another limiting factor – water. Cornish mines were (and still are) very wet. It was Newcomen's invention of a workable steam engine in the early eighteenth century, later developed by Watt in the 1770s and improved by Trevithick and other Cornish engineers, which made the pumping out of deep mines possible and heralded the heyday of Cornish mining. Some of these giant engines, with cylinders over 7 feet in diameter, worked day in and day out for over 50 years. The last mine pumping engine ceased working at South Crofty, Camborne, in 1955, having first started work in another mine in 1854 – over 100 years of service! The biggest steam engines the world has ever seen, with cylinders a staggering 12 feet in diameter, were made in Cornwall. As well as pumping water out of the mines, steam was used for driving the hoisting machinery in the shafts, and the whole county would have resounded to the roar of the Cornish stamp machines which were used to crush the ore.

Underground the work was arduous. Miners had to break up some of the hardest rock the Earth can provide with only hand-held tools and gunpowder. There was little ventilation, and the miner worked by the light of a candle stuck to his felt hat with a lump of clay. Pay was varied, job-security non-existant and accidents were common. Cornish miners were very superstitious and women were never permitted to work underground – that would have been bad luck. Much of the ore dressing on the surface was done by women and children. As the boys became older many of them went underground to work with their fathers and uncles. It was literally a sub-culture, with customs, working practices and a language of its own. Lunch-break is still *croust* in Cornwall, and no one calls a spade a shovel – the Cornish call them *banjos*.

84 Multi-hand boring of rock.

85 The Dolcoath man-engine. The men jumped on and off as the ladders moved up and down.

ROCK SOLID

THE CHANGING FORTUNES OF THE MINES

Cornwall's history has grown from the richness of its geology, yet the fortunes of those who mine those riches have been dictated by politics and world economics as much as by the assay of the ores hewn from the rocks.

World prices for metals have long been dictated by supply and demand – the rarer the metal the higher the price. Twice in its early heyday Cornwall's copper mining was hit by competition – first, in 1768, when copper was discovered in Parys mountain, Anglesey and again in 1866 when competition came from copper mines in North America. The tin market suffered similar collapses in the 1870s and 1890s, and by 1900 only a few of the larger and richer tin mines had survived, due to a combination of economic factors and to the best lodes having been exhausted. By 1930 only three mines were in production, to be reduced to two, South Crofty and Geevor, by 1946. Demand for tin, tungsten and copper during both world wars marked revivals of activity, but the peace of the late 1940s heralded what looked like being the final death throes for Cornish tin.

Then, in the 1960s metal prices on world markets started to soar. This attracted investment from multinational mining finance houses which, together with sophisticated underground mechanisation and the ability to smelt metals from ores of a much lower grade, caused another boom in Cornish tin, and the re-opening of several mines – one of the most important being Wheal Jane. In 1984, the production of tin in Cornwall was the highest since 1915. Things were looking promising.

Then in 1985 the price of tin on

86 South Crofty is Cornwall's last working mine. The modern winding gear stands alongside the old granite engine house.

87 A geologist taking a sample from a mineral lode. The sample will be chemically analysed at the surface. The ore is in finely divided particles dispersed in the lode.

world markets collapsed from over £6000 a tonne to £4000. Geevor closed. An interest-free loan of £15 million from the government kept South Crofty and Wheal Jane going until March 1991 when Wheal Jane shut down. Appeals for funds to keep the pumps working, until such a time as demand for tin copper and zinc increased, fell on deaf ears. So the mine, with all its sophisticated machinery, now lies soaked in corrosive waters, still and silent forever. At the time of writing, South Crofty is still working, hanging on by the skin of its teeth, by virtue of being sited on what has always been the most richly mineralised and profitable part of Cornwall. Should South Crofty die (the only mine to be worked continuously since the 1670s) it will mean the end of a 3000-year era for Cornwall, and the end of serious hard-rock mining in Britain.

88 Froth floatation is a method of separating the metal ore from other, unwanted, minerals. The metal ore particles attach to the bubbles, but the other minerals sink.

THE MINING HERITAGE

Cornwall's mines have left an indelible mark on the landscape and on the lives of Cornish people. Many water courses are polluted by old tailings The stream in fig 89 has turned red with suspension of iron ore. The ground is riddled with shafts and tunnels, some sealed, others not. Several mining operations have opened as tourist attractions, such as *Cornish Engines* in Camborne, and *Poldark Mine* near Helston. Finally, something that will warm the hearts of people for many years to come is the Cornish pasty, the traditional food of the miner. This is a complete meal of meat and root vegetables enclosed in a

90 Old mine shaft.

pastry case which keeps hot for hours, and its traditional rope-shaped crust is designed not for eating, but for holding the pasty with dirty fingers!

89 Pollution has turned the water of this river red.

ROCK SOLID

CHINA CLAY

Although granite is extremely tough and resistant to weathering, heat and prolific amounts of water can cause the feldspar crystals to break down and turn into the clay **kaolin**. This happened to Cornish granite during its cooling and hydrothermal mineralisation; and again much later, between 50 and 25 million years ago, when the granite was exposed to weathering in a subtropical climate in the geological era known as the **Tertiary**. The kaolin stays where it was formed, still mixed up with the other granite minerals, quartz and mica, but it is no longer a hard rock – more of a crunchy mush. Large areas of Cornish granite are kaolinised, particularly in the St Austell area which still produces 2.5 million tonnes of kaolin every year for use as china clay.

To extract the kaolin a powerful jet of water (called a **monitor**) is fired at the quarry face. A white slurry runs away from the face to a processing plant which separates the clay from the quartz and mica. It is these 'waste' minerals which form the tall, white conical waste tips that are such a prominent feature of the eastern part of the Cornish landscape.

China clay is used, as the name implies, to make posh white bone china, but this is not its main use. The bulk of it goes into the manufacture of paper. The pages of this book contain between 30 to 40 per cent china clay.

HOT ROCK

Cornish tin mines are not only wet, but very warm, often well in excess of 30°C. Even when winter snow lies above ground, the miners sweat away in steamy sauna bath conditions. The deeper you go, the hotter it gets. But where does all the heat come from, so far away from the warmth of the Sun?

The answer lies in the rocks – in minerals which contain **radioactive** elements. The atomic nuclei of these elements break up into smaller nuclei, giving out heat energy as they do so. Cornish granites are rich in these minerals, the most common being a particularly attractive uranium mineral, torbernite (fig 92). Experiments were carried out to exploit this heat, as **geothermal energy** in Cornwall's **Hot Dry Rock** project. Holes were drilled deep into the granite and cold water pumped down, where it circulated through natural cracks in the rock and heated up. It was then pumped back to the surface. This might sound more like hot *wet* rock – it is called *dry* because water needs to be injected artificially into the rock. In some other areas, such as Bath, warm springs reach the surface without human intervention.

In Rosemanowes, Cornwall, the rocks were drilled to a depth of 2.6 kilometres, and a temperature of 100°C was reached. It has been estimated that if drilling had been continued down to 6 kilometres, rock temperatures of around 200°C would have been found. This would be hot enough to generate

91   A china clay pit near St Austell.

92   The green uranium mica, torbernite.

electricity efficiently and economically, but sadly the money for the project ran out. However, this renewable energy resource is still there, and desk-top research into Britain's geothermal energy potential continues. So one fine day, we could find some of our electricity generated using the Earth's own, environmentally safe, nuclear power.

The decay of uranium also produces **radon** gas which is itself radioactive and can be harmful to people. Radon constantly seeps up from the granite, and if a pretty cottage is built both of granite and on top of granite, it could pose a health-hazard. High-risk houses have to be kept well-ventilated to stop the concentration of radon from rising, and all underground workings have to be monitored very regularly.

94 Levels of the radioactive gas radon have to be checked regularly in Cornish mines.

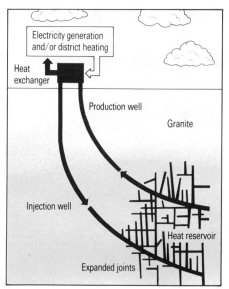

95 How the water is circulated to capture the heat energy.

93 Underground Cornwall is very hot – these miners are working in temperatures well in excess of 30°C.

96 The site of the Hot Dry Rock project in Rosemanowes Quarry, Cornwall.

ROCK SOLID

ROMANCE AND REALITY

The Dorset town of Lyme Regis is a Mecca for fossil collectors. For at least two hundred years the plethora of fossils washed out of the cliffs has fascinated amateurs, professional collectors and scientists alike. The distinctive air-force blue beds of rock were laid down in the sea at the same time as the dinosaurs were starting to take over the land, in a period of geological history known as the **Jurassic** (between 145 and 208 million years ago). The mudstones and limestones that make up the Lyme cliffs are known as the **Lias**, probably from an old French word meaning 'layered', for that is just what they are – layers of sedimentary rock. Lias also refers to the time when the blue-coloured rocks were laid down, over several million years. You can see when this stage ended at the top of the cliffs. Here the rock type changes abruptly to golden sandstone, laid down in the sea during the **Cretaceous** period, towards the end of the dinosaurs' reign.

The wonderment of finding a fossil that has not seen the light of day for millions of years, and never before been gazed upon by human eye, is an emotion that always touches the most hardened of hearts and stirs real passion in fossil enthusiasts. This romantic side of geology was illustrated by the author John Fowles in his novel *The French Lieutenant's Woman*. The hero, Charles, is a typical Victorian gentleman geologist

97 The cliffs of Lyme Bay.

98 The heart-shaped fossil that touched the heart of the hero in *The French Lieutenant's Woman*, the fossil sea urchin, *Micraster*.

fascinated by the then new theories of evolution proposed by Charles Darwin. On a fossiling expedition in the picturesque undercliff to the west of Lyme, Charles comes across Sarah, the 'French lieutenant's woman' – she has something he finds irresistably attractive:

She delved into her pockets and presented to him, one in each hand, two excellent Micraster tests [fossil sea urchins].

'Will you not take them?'

She wore no gloves, and their fingers touched. He examined the two tests; but he thought only of the touch of those cold fingers.

'I am most grateful. They are in excellent condition.'

'They are what you seek?'

'Yes indeed.'

'They were once marine shells? . . .'

Charles goes on to explain that they were indeed once marine shells from the younger Cretaceous rocks, which had slipped down in the landslip that forms the undercliff. A liaison between the two is born, but a liaison strewn with rocks metaphorical as well as geological.

The fossils of Lyme Regis were made famous around the start of the nineteenth century by the Anning family: Richard Anning, his wife Mary, son Joseph, and most famous of all, his daughter, also called Mary. Richard Anning died in 1810 when Mary was only 11. She and her brother continued to collect and sell fossils to support themselves and their mother. A market for fossils was growing: first to the wealthy, visiting Lyme and the seaside between the London and Bath 'seasons'; and then to scientists, who were beginning to take a serious interest in such things.

The younger Mary Anning was a remarkable person of high intelligence and with a versatile range of skills. Women palaeontologists are few and far between even today, yet Mary, with no formal education, taught herself palaeontology and zoology (one needs to study the anatomy of living animals in order to interpret and understand the petrified remains of those of the past). She traded and corresponded with the most informed scientists of her day, and she became an astute business woman, identifying and pricing her fossil finds with care and acumen.

Mary and her brother were responsible for the discovery and excavation of the first near-complete **ichthyosaur** in 1812. They sold it to the lord of the manor for £23, but it eventually found its way to the Natural History Museum in London where it still resides. The site of the Annings' first house in Lyme Regis is now occupied by the *Lyme Regis Philpot Museum*.

99 A posthumous portrait of Mary Anning by B J M Donne.

100 The skull of one of Mary Anning's ichthyosaurs, now in The Natural History Museum.

ROCK SOLID

ANCIENT SEA MONSTERS

The discoveries made by the Anning family captured the Georgian and Victorian imagination and led to many reconstructions, some scientific, others more fanciful, such as the one in fig 101, of fearful monsters battling it out in a turbulent sea. In fact, the vast majority of ichthyosaurs and **plesiosaurs** were almost certainly more peaceful creatures which fed on fish, squids and squid-like animals called belemnites.

Ichthyosaur fossils are more common than plesiosaurs, and so more is known about them. The name ichthyosaur means 'fish reptile' and their body plans do have characteristics of both fish and reptiles. They bear a strong resemblance to modern dolphins and whales, which are neither fish nor reptile, but mammals – mammals whose ancestors had lived on land but became adapted to aquatic life where there was a plentiful supply of food with fewer predators. Similarly, the ichthyosaurs must have had a land-dwelling ancestor which took to the seas to occupy the same niche, and developed the body plan best adapted to that niche. So, a similar appearance need not imply a common ancestor, but is likely to indicate a similar lifestyle. Ichthyosaurs, like dolphins, were entirely sea-dwelling, but plesiosaurs may well have been able to haul themselves onto the land with their paddles. They probably swam more slowly than icthyosaurs.

102 A modern reconstruction of the Liassic seas of Lyme Regis with ichthyosaurs and belemnites (squid-like molluscs).

101 An imaginary battle between ichthyosaurs and plesiosaurs from Thomas Hawkins' *Great Sea Dragons*, published in 1840.

DORSET'S DINOSAURS

The marine reptiles of Lyme Regis are not dinosaurs. Dinosaurs were a distinct and separate group of land-dwelling reptiles, distinguished from other reptiles by their limbs which went straight down from the body rather than being splayed out to the side like crocodiles and lizards. One species of dinosaur has been found in the rocks of Lyme Regis – *Scelidosaurus*, a plant-eating dinosaur with a beak-like mouth and rows of horny spikes along its back. Fig 103 is a reconstruction of a specimen found in 1985 by a group of professional collectors. It was found in marine sedimentary rocks, but was evidently a land-dwelling animal; it must have been washed down into the sea by a river.

Fossils are not always found in rocks of the environment in which they lived (indeed, sometimes it was that environment that proved so hostile as to kill them). This marine grave for the *Scelidosaurus* had an added bonus: fig 104 shows fossilised skin, with scales clearly visible, preserved in the rock. The preservation of soft parts is extremely rare, particularly with dinosaurs, but the lack of oxygen in the mud at the bottom of the Liassic sea, would have made the process of decay very slow, giving time for this dragon-like glimpse of the past to be preserved.

In east Dorset, dinosaur footprints are often found in younger rocks laid down in shallow water towards the end of the Jurassic period. These tracks were made by a meat-eating *Megalosaurus* about 145 million years ago. Such footprints are a delight to scientists and dinosaur lovers, but a real nuisance to quarrymen. The great weight of the animal disturbed the sediment (laid down in a shallow coastal lagoon), which is now consolidated into rock, to a considerable depth, rendering a great deal of stone unsuitable for masonary.

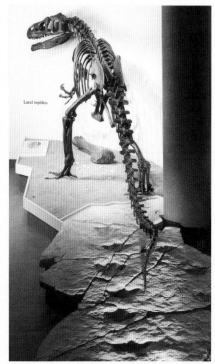

105 Dinosaur footprints in Purbeck stone, with a skeleton of the type of dinosaur that probably made them, *Megalosaurus*.

103 A model of *Scelidosaurus*. 104(inset) Fossilised *Scelidosaurus* skin. Fossils of soft parts of animals are extremely rare.

ROCK SOLID

106 An ammonite, *Parkinsonia dorsetensis*, polished to show the frilly suture lines.

AMMONITES

The elegant coiled shells of ammonites are probably the most familiar fossils to the layman, and are common on a great number of Dorset's beaches, particularly the stretches on either side of Lyme Regis. They are an extinct group of molluscs, distantly related to snails, and more closely related to squids, octopuses and the living nautilus. Like the nautilus, but unlike snails, the shell of the ammonite was divided into chambers by a series of septa. The septa of the nautilus are smooth and curved, but those of the ammonite were complex and highly frilled at their edges (fig 106), so much so that the join between the septa

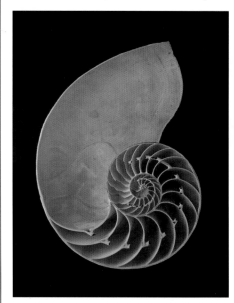

107 Cross-section of a modern *Nautilus*, showing the septa which divide the shell.

and shell (the **sutures**) are reminiscent of the fine Honiton lace that used to be made a little way down the coast in Devon. The animal lived in the outermost chamber. The inner ones were filled with fluid and gas, the proportions of which could be varied so that the ammonite could rise or fall in the sea at will, rather like a submarine.

The purpose of the frilly sutures is uncertain – they would not have been visible on the outside of the shell in life, ruling out any decorative function. It was once thought that they could have strengthened the shell, enabling it to withstand the pressure changes of moving between different depths of water. The complexity seems certain to have been a means of increasing the surface area of each septum – but for what purpose, no-one, as yet, knows.

No part of the Jurassic and Cretaceous seas seem to have been without ammonites. They are found in almost all marine sedimentary rocks of those ages. Ammonites come in all shapes and sizes: some are large, others small, some like coiled snakes, others fat with outer whorls completely enveloping the inner ones, some slim and disc-shaped. There are knobbly ones, ribbed ones and some with bizzare spikes – the variety is almost endless. Some shapes may have developed to adapt to different modes of life and diets – for example, a slim disc-shaped ammonite may have swum swiftly through the water pursuing its prey, whereas a fat one may have sat motionless in the water and waited for its

108 An artist's impression of ammonites and belemnites swimming in the *Jurassic* sea 200 million years ago.

ROCK SOLID

109 Mr & Mrs Ammonite – the larger *Liparoceras* is the female, the smaller *Aegoceras* the male.

dinner to come to it, in the form of plankton.

Some of the different ammonite shapes have now been attributed to sex differences – the females were bigger than the males, and presumably dominant. Some people (including the present author) think this is why they were so successful and thrived for 150 million years! All ammonites, however, died out at the same time as the dinosaurs, 65 million years ago. No one theory explains adequately why this mass

extinction on land and in the seas took place. The impact of a meteorite has been blamed, huge volcanic eruptions in India may have caused global cooling, or changes in sea levels may have disrupted ecosystems in the sea and climatic patterns on land; more likely, all three factors were involved.

The sheer beauty of ammonites makes them valued and enjoyed as ornamental objects. But they are very useful to science, and to the oil industry too, because they are used in dating rocks. As they evolved very rapidly, ammonites of markedly different appearance occur in beds of rock separated by short time spans (short in geological terms, that is!). The whole of the Jurassic has been divided into time zones on the basis of particular species of ammonite being found only within each individual zone. Fig 110 shows the zone ammonites for a typical section of cliff at Lyme Regis, each one representing a period of roughly 700 000 years.

Ammonites were entirely marine animals, so their presence indicates that a sedimentary rock was laid down in the sea (as opposed to a lake, river or delta). The environment in which a rock was laid down can give important clues to the geologist prospecting for oil.

FOSSIL COLLECTING – MONEY AND MORALS

Lyme Regis is home to two large and prosperous fossil shops as well as two museums, *The Lyme Regis (Philpot) Museum*

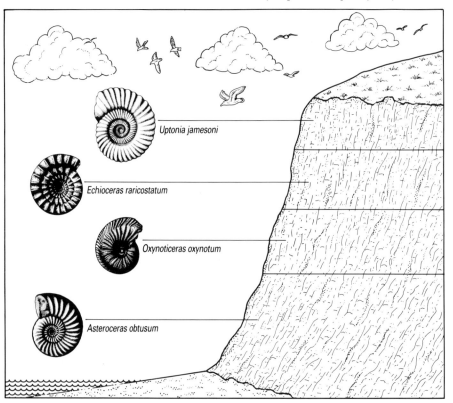

Uptonia jamesoni

Echioceras raricostatum

Oxynoticeras oxynotum

Asteroceras obtusum

110 Zone ammonites can be used to date sequences of rocks.

and *Dinosaurland*. It is also home to several professional collectors – those who make a living by collecting and selling fossils. Some people feel uneasy about this. Indeed some feel very strongly that professional collectors and amateurs alike should be banned. This view was expressed in *The Independent* of 2 June 1991 which spoke of '. . . *fossil hunting Klondykers systematically vacuum cleaning all fossil-rich sites. . .*' Undoubtedly there is some truth in this – professional collectors have taken over quarries and sold all the fossils to shops. All too often, the fossils are sold without a proper label describing exactly where the fossil was found and what it is. This entirely removes the fossil's value as a scientific object. Sometimes fossils are very highly polished, or part of the fossil removed to enhance its beauty as a decorative object and this too can destroy scientifically valuable parts.

But there is another side to the argument. The cliffs of Lyme Bay are continually being eroded by the sea and rain. If fossils washed out of the cliffs are not collected, they will be destroyed by the waves and go back into the sea from where they came 200 million years ago.

Many professional collectors have worked closely with museums and scientists to make some very valuable material (valuable in both scientific and monetary terms) available for research, and for exhibition where it can be enjoyed by the public. The *Scelidosaurus* on p 53 is a good example of this. Figs 111 and 112 show two other examples

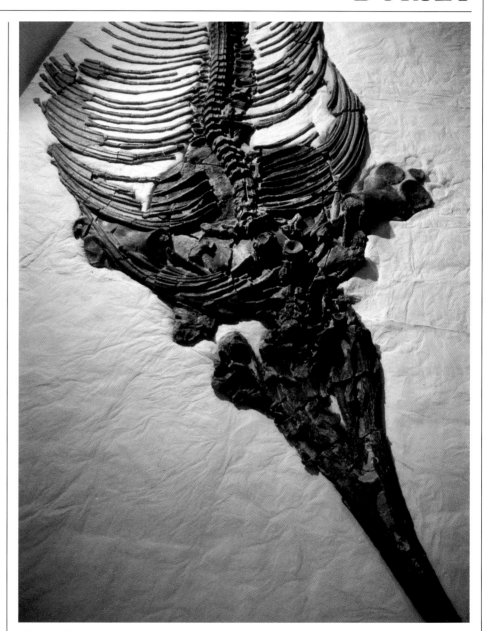

111   The Charmouth ichthyosaur – the largest complete specimen ever found.

of this type of cooperation. Fig 111 is an ichthyosaur found in the cliffs at Charmouth by the professional collector David Sole. This is not how it would have looked when it was found – it is all too easy to imagine stumbling on a complete ichthyosaur neatly laid out on the sea shore just as they look in museums, but it just doesn't happen like that. Spotting them embedded in the rock is an art in itself, as is the long

process of extraction, and then the fossil has to be professionally cleaned and prepared. This specimen turned out to be the largest ichthyosaur ever found and was sold to the City of Bristol Museum. The smaller ichthyosaur (fig 112) was also sold to the Bristol Museum by another well-known professional collector, Peter Langham, who found it in rocks of the same, Liassic, age in north Somerset. It is important scientifically because it is a female carrying an embryo, proving

conclusively that ichthyosaurs did not lay eggs like other reptiles, but gave birth to live young, just as modern dolphins do.

Amateur collecting the responsible way

If you are fed up with sandcastles and shrimping, and it is not quite time for that cream tea, what better way to spend time on the beach than saving fossils from the wrath of the waves, and to wonder and delight at being the first human ever to see or to handle your finds. There are a few basic rules:

• If you can, try to go fossiling on a guided tour with an expert. Local museums will have details. In Lyme, the Charmouth Heritage Coast Centre run regular fossiling trips (fig 115)

• Most guide books tell you that you

112 The Watchet pregnant ichthyosaur with (inset) her embryo.

113 There is no indication of where this ammonite came from or what species it is, and it has been over-polished. It could hardly be described as a bargain!

will need a hammer. They're helpful but not absolutely necessary – in Lyme Bay, at least, you can pick up little ammonites and their cousins, the belemnites, just lying on the beach. If you do hammer, do it with care, and away from other people.

• Safety – **don't bash away at cliffs**. Collectors (amateur and professional) have been killed by cliff falls, and the sea does a much better job of getting fossils out of the cliffs anyway.

• If you find something you cannot identify, take it to an expert or the local museum. Take some of the rock you found it with, and an **exact** description of where you found it.

• **Don't be greedy and pinch the lot.**

• Look after your finds and treasure them. Write down exactly where you found each one, and store them carefully with numbers fixed on each fossil corresponding to the labels you have written.

If you have a fruitless search, go and enjoy the fossils in the museums and shops. If you buy one, make sure it has a proper label saying exactly what it is and precisely where it came from – your fossil will then have its full scientific value to back up the monetary value.

116 This letter from Mary Anning, written in 1831 to an academic geologist, shows that she saw the value of the rare fish she is describing in both scientific and financial terms (she is asking £50 – a considerable price in those days). She went to a great deal of trouble to identify the specimen and to make it available to scientists.

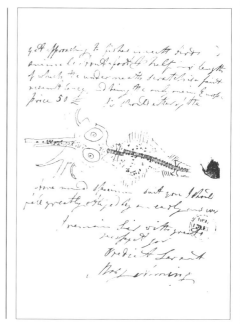

114 A clearly labelled specimen from a private collection.

115 Richard Edmonds, the Warden of Charmouth Heritage Coast Centre, leading a group of children on a fossiling expedition.

117 Oil seeps out of these cliffs at Osmington Mills, Dorset.

THE WYTCH FARM OILFIELD

The Wytch Farm oilfield underlies the northern sector of the Isle of Purbeck and much of Poole Harbour. It is the largest on-shore oilfield in Western Europe and the sixth largest in the UK (the five bigger ones are entirely under the sea). It produces 60 000 barrels of oil and 10 million cubic feet of gas every day. Oil has been vital to Britain's post-war economy – indeed some would argue that oil is Britain's post-war economy. Prosperity from oil is due to our geological history – the time we spent submerged under the warm Jurassic seas 200 million years ago.

Oil is formed from the remains of the soft parts of marine organisms which became buried in mud in deep

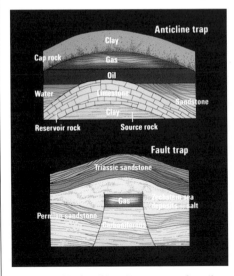

118 Two kinds of 'trap' structures for oil and gas – *top*, a fold and *bottom*, faults.

water where there is very little oxygen, so the normal processes of decay are incomplete. These conditions prevailed at the bottom of Dorset's ancient Liassic sea. As the organic remains became buried by sucessive layers of sediment, the pressure and heat slowly transformed them into the cocktail of chemicals which makes oil.

Oil, being light, tends to move upwards in the rocks. Sometimes it escapes at the surface as a seepage (fig 117). But, if it meets a permeable rock, such as a sandstone, it will be soaked up like water in a sponge. If that rock is overlain by a layer of impermeable rock, such as clay, and if the oil meets a structure which stops it migrating sideways, such as a fold or a fault, it will be confined on all sides, in what is called a **trap** (fig 118). Oilmen like nothing better than to release that trapped oil from its underground prison, which they call a **reservoir**.

In the 1970s the oil companies began to look at Dorset. Knowing that the Liassic rocks were a likely source for oil, that there were plenty of suitable sandstones in the vicinity that could act as reservoirs soaked with oil, and that all the rocks in the county were pushed into folds and faults at the time the Alps were being formed between 50 and 20 million years ago, Dorset looked promising.

The Wytch Farm oilfield turned out to contain not just one, but two reservoirs, one beneath the other (fig 119). There was concern that Dorset would be transformed into Dallas, and that Thomas Hardy would turn in his grave. But, when full production was reached in 1990, the well sites and nodding donkeys (fig 120) were so efficiently concealed that the only way the public could see them was with a pair of binoculars from the top of the nearby ruins of Corfe Castle (fig 121). When the oil runs out, all the sites and installations will be restored to their former rural glory, leaving not so much as a stetson behind.

120 Nodding donkeys at one of the Wytch Farm wellsites.

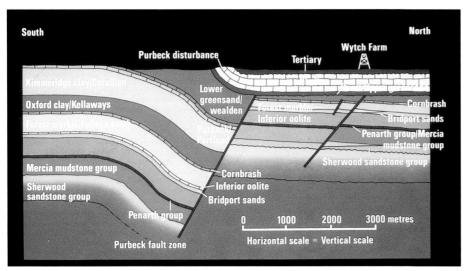

119 A cross-section of the rocks under the Isle of Purbeck. Oil was found in the Bridport sands and the Sherwood sandstone, in fault traps.

121 The picturesque village of Corfe Castle.

ROCK SOLID

A WEALTH OF GEOLOGY

London and Paris have much in common, and the likeness runs considerably deeper than fashion, the arts and fast-living. Both are built on clays and sands contained within a giant basin of chalk (fig 122).

Chalk is a soft, white form of limestone made up of the skeletons of countless microscopic organisms (fig 123). These organisms lived in a warm

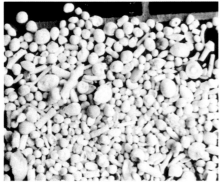

123 Chalk under the microscope – a myriad of microfossils. Blackboard chalk is not the same; it is made from the mineral gypsum.

sea which extended from Scotland to the South of France between 97 and 65 million years ago. Over that considerable period of time, the skeletons fell like a long snow fall to the sea bed and formed a 500-metre thick, white blanket of sediment. The sediment hardened to chalk and by 60 million years ago much of it had risen to become dry land as global sea levels dropped. Life on Earth was in the throes of catastrophic change – the dinosaurs had died out, as had the ammonites and many other species in the sea.

A final gentle push

The chalk would have stayed as a thick flat white blanket overlain with younger clays and sands, were it not for the fact that around 50 million years ago, Africa started to shove its way into Europe. The rocks of Switzerland, northern Italy and the South of France were pushed together in a contortion of folds, one on top of the other, to form the Alps.

Further north, away from the main action, the rocks were pushed into more gentle folds: down to form the Paris Basin; up over Normandy; down under the channel; up again in the White Cliffs of Dover and the South Downs; high over the Weald of Sussex and Surrey (this dome of chalk has since eroded away) and back down to become the North Downs; down under London then up again to form the Chilterns.

Chalk absorbs water. Because of the basin structure of London's chalk, rain falling on the Chilterns and North Downs soaks into the ground and down into the chalk underneath the capital. The London Clay on top, and a layer of clay underneath act to seal the water into a sizeable underground reservoir, or **aquifer**. In earlier centuries, wells dug through the London Clay into the chalk resulted in clean fresh water gushing to the surface. As London's population grew and industry developed, so much water was taken out of the aquifer that water no longer rose of its own accord and needed to be pumped up the wells.

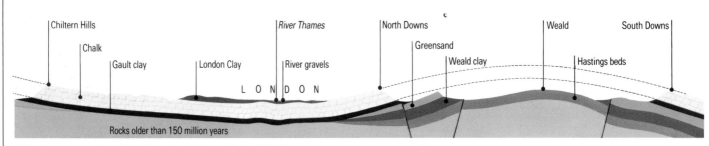

122 A cross-section of the London Basin and the Weald.

Gradually, alternative sources of water had to be found, from rivers and artificial, surface, reservoirs built on the edge of the metropolis.

But in recent years, with the decline in traditional water-using industries, such as brewing, printing and papermaking, water levels under central London have started rising again, by a staggering 1 metre a year. Foundations are threatened and the London Underground is at serious risk of flooding. Thames Water are now investigating the possibilities of drilling new wells, to exploit once again the aquifer. The water tables have turned, so to speak.

124 Monitoring the quality of water from a borehole on the outskirts of the London Basin. Most aquifer water is very pure, because the rock acts as a natural filter, but impurities can result from farm waste, or from rubbish tipped down old well shafts.

125 London Underground tunnel: the Victoria line under construction, 1966. The London Clay which lies on top of the chalk is easy to tunnel through, being soft yet holding its shape rather than collapsing. So, the sub-surface London has become a labyrinth of tube tunnels and sewers.

ROCK SOLID

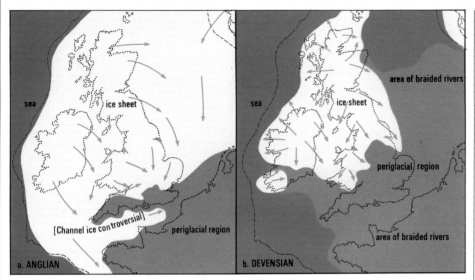

126 The extent of the ice-sheets in the Anglian (400 000 years ago) and Devensian (50 000 years ago) glaciations.

OLD FATHER THAMES

For the past two million years an **Ice Age** has waxed and waned over Europe, putting the final touches to our landscape in the process. At least four advances of thick ice-sheets have left their mark, with warmer **interglacial** periods in between. Fig 126 shows the extent of the ice-sheet that reached the furthest south, the **Anglian Glaciation**, about 400 000 years ago; and the most recent ice-sheet, the **Devensian**, which melted 10 000 years ago. The site of the City of London itself has never been covered by ice, but during the glacial advances it would have been a bitterly cold tundra populated by mammoths, woolly rhinoceroses and reindeer. During the interglacial periods

127 Trafalgar Square during the Ipswichian interglacial, about 100 000 years ago.

the climate would have been quite pleasant and temperate, at times warmer than today.

Fig 127 shows what Trafalgar Square would have looked like during one of these interglacials – populated by straight-tusked elephants, hippos, lions, deer, bears and bison; with not a single pigeon in sight. This reconstruction is not fanciful, but based on fossil evidence unearthed during digging for foundations. A city such as London can offer more geological information than some rural areas, because of the quantity of holes that are dug for construction purposes. To dig such holes specifically for geologising would be prohibitively expensive.

Fig 128 shows the course of the River Thames today, compared with half a million years ago, when it flowed further north through St Albans and Essex. It reached the sea somewhere near Harwich. But when the Anglian ice-sheet advanced, the river's course was blocked by a wall of ice, forcing it to take a turn towards a

DEATH OF THE DINOSAURS

No-one knows for sure what caused the **mass extinction** of dinosaurs and many other creatures 65 million years ago. A large meteorite is known to have crashed into the Earth about this time, and vast quantities of dust and gases were being spewed out of volcanoes in India – both have been blamed. But it could be that the sea level fell, insidiously contributing to the extinctions by reducing the area of shallow shelf sea (the favourite haunt of the ammonites). This would have disrupted continental climates and might have caused a global temperature change to which the dinosaurs could not adapt. We shall probably never know for certain what happened, not least because the evidence for this change in sea level in the London Basin is negative rather than positive. Most of the top layers of chalk, including any rocks laid down at the time of the extinction, were eroded away while they were exposed as land. The next rocks to be laid down came about 10 million years later, when southern Britain was again inundated, this time by a muddy sea which laid down the London Clay. So the mass extinction has left its mark in London only as a gap in the rock record.

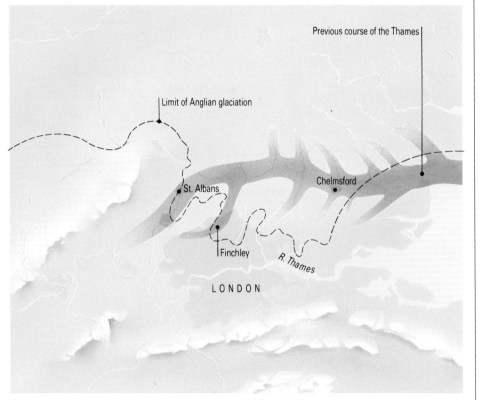

128 The Thames once took a more northerly course.

ROCK SOLID

more southerly route which, give or take a few changes in meanders, it has adopted ever since. The old course is marked by deposits of gravel, now the Southeast's most important source of aggregate for construction.

During glaciations, so much water was locked up in the ice-caps that sea levels fell dramatically. This caused the Thames to cut a new and deeper channel into its valley, leaving remnants of its old flood plain as terraces stranded on the old valley sides. Three stages of terraces have been traced either side of the Thames in London, one of the most obvious being the steep hill that leads from the Embankment north to the Strand.

In the earliest set of Thames river terraces at Swanscombe in Kent, were found not only flint tools, but the oldest human remains to have been discovered in Britain – three skull fragments. Originally dubbed **Swanscombe Man**, this earliest proven inhabitant of Britain is now believed to have been a woman. She is thought to be about 250 000 years old and an early **Neanderthal**, one of a small community that hunted game and gathered seeds and berries during one of the interglacial periods. She is unlikely to have been one of the direct ancestors of Europeans, who are now thought to have been derived from the **Cro Magnons** who migrated to Europe from Africa and settled in Britain about 35 000 years ago. They may have interbred with the indigenous Neanderthal population, but there is no definite proof of this.

Sadly, London's long chapter of geological and human history will one day come to an end. The whole of Southern Britain is subsiding, and sea levels appear to be rising. This rise is because of the **greenhouse effect** which is caused by increased carbon dioxide levels in the atmosphere emitted during the burning of our older geological riches: coal and oil. Attempts have been made to hold back the sea – by King Canute's polite, but futile request to the tide in the tenth century; and by the building of the Thames Barrier in our own, twentieth, century.

Britain has been bobbing up and down, in and out of the waves, throughout geological history, and it is doubtful that humankind can hold the sea away forever. But, it is possible that the next glacial advance could be with us sooner rather than later (in a couple of thousand years maybe), causing a drop in sea levels to counteract the sinking land and global warming.

129 A deer hunt on the banks of the river at Swanscombe 250 000 years ago. A wooden spear has brought down a specimen of the extinct Clacton fallow deer *Dama dama*.

STREETS PAVED WITH ROCKS

The chalk, sands and clay of the London Basin do not provide any hard stone for building houses, so when the Romans arrived and decided to build a wall round the City, they had to bring stone in from the Weald of Kent (fig 130). When William the Conqueror came in 1066, he brought his own stone with him from Caen, in France, to build the central keep of the Tower of London.

As London grew, building stone from an increasingly wide area was brought in. Inigo Jones, the architect and surveyor to James I of England (James VI of Scotland), introduced Portland stone to London from Dorset to build the Banqueting Hall in Whitehall.

130 Part of the London Wall, constructed in Roman times from stone brought in from the Weald of Kent.

131 The White Tower of the Tower of London, built of Caen Stone from Normandy.

ROCK SOLID

After the great fire of London, in 1666, Christopher Wren used Portland stone to rebuild St Paul's Cathedral and established it as one of London's most important building stones, and one which has added elegance to London's character.

Many modern buildings have a steel and concrete structure, which is then faced with sheets of polished stone only three or four centimetres thick. A wondrous variety of rocks are used in this way – rocks from all round the world: sedimentary, igneous and metamorphic, rocks of all ages, each with a story to tell. It is worth pausing in any London street (or the streets of any town) to admire a magnificent urban exhibition of geology.

132 *(left)* The West front of St Paul's Cathedral, built with all kinds of rocks from all kinds of places.
(a) Black and white marble paving.
(b) Steps made of a hard-wearing igneous rock called diorite.
(c) Blocks of Swedish limestone with fossils, set in white Italian marble.
(d) Blocks of grey granite from Cornwall.
(e) Blocks of Purbeck limestone, some with the fossils of freshwater snails.
(f) The statue of Queen Anne is carved from white Italian marble.
(g) Bollards of pink Shap granite from the Lake District.

133 Larvikite, an igneous rock from Norway, is so common in British public houses that it has been dubbed 'Publichouseite'. The sparkle and iridescence is due to the play of light on planes within the crystals of feldspar. There are two varieties of Larvikite, blue pearl and emerald pearl. This is the blue variety.

136 Shap granite from the Lake District cooled in two stages. The first was very deep in the crust when the large salmon-pink crystals of feldspar formed. The magma, with the crystals, was then forced up to a higher level in the crust, where the rest of the liquid crystallised to form the finer mesh of crystals in the groundmass. The colour is due to the presence of small quantities of iron.

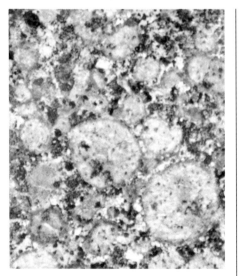

134 Baltic Brown granite from Finland is very popular in modern high streets.

135 The floor of the St Andrew's Chapel in Westminster Cathedral is laid with folded marble, reputedly from Scotland.

137 There are rocks in every high street! This Green slate from the Lake District was formed from volcanic ash which fell into the ancient *Iapetus* Ocean. It is fine-grained and smooth, and so can take delicate carving.

ROCK SOLID

One of the few rock types not represented in Britain is **kimberlite**, the volcanic rock in which diamonds are found. Yet, 80 per cent of the world's diamonds are traded in London. Diamonds are far more than a girl's best friend – their industrial applications are numerous due to their hardness, their lack of chemical activity and their ability to conduct heat. They have long been used for cutting and grinding; recently, the Voyager spacecraft were fitted with a small diamond window (about one centimetre across) which was strong, optically perfect, and able to withstand any acid atmospheres it might encounter. Diamond has myriad applications in electronics; indeed, any country with a sophisticated engineering and electronics industry must have access to diamonds in order to survive and develop. Again, it is the rocks that determine our prosperity.

138 Diamond crystals in kimberlite from Kimberley, South Africa (top right), and Siberia (centre), and in a beach conglomerate from Namaqualand (top left). Diamonds from all over the world are traded in London.

GLOSSARY & CREDITS

Ammonites Extinct molluscs with coiled shells, distantly related to the modern nautilus.

Aquifer A permeable underground body of rock, soaked with water.

Basalt A black volcanic rock.

Belemnite An extinct mollusc, related to the squid. Usually, only the hard cigar-shaped guard, which protected the fragile shell, is preserved.

Cleavage The tendency of a rock or mineral to split along flat planes. The best-known example is slate.

Coal seam Coal occurs in layers, or seams, anything from a few centimetres to a few metres thick.

Convection currents The motion set up in fluids when the hotter parts flow to the colder. Such currents occur in the Earth's mantle.

Crust The thin, outermost layer of the Earth, made of solid rock. Compared to the Earth as a whole, it is thinner than the skin on an apple.

Dyke A sheet of igneous rock, usually an underground off-shoot of a volcano, which cuts through layers of older rock.

Epicentre The point on the ground directly above the focus of an earthquake.

Extrusive A rock formed from material extruded from a volcano.

Fault A fracture in rocks along which there has been movement.

Feldspar A common mineral consisting of silicon, oxygen and aluminium, with varying amounts of calcium, sodium and potassium. It is one of the main constituents of granite where it occurs as white or pink crystals.

Focus The point at which an earthquake actually happens, along a fault underground.

Granite A speckly coarse-grained igneous rock which formed from magma which cooled and crystallised inside the Earth's crust. Its overall colour is usually grey or pink.

Hotspot An isolated region of the Earth's mantle, sufficiently hot to allow the rocks to melt.

Hydrothermal mineralisation The deposition of minerals from warm or hot solutions circulating in the crust.

Iapetus An ancient ocean that once divided northern Britain from southern Britain.

Igneous rocks Those formed by cooling and solidification from a molten magma.

Interglacial A warmer period during an ice age, in between advances of the ice sheet.

Joints Cracks in rocks.

Lava Molten rock (magma) extruded from a volcano.

Lias, Liassic The name given to both the rocks and to the epoch (210-178 million years ago) at the start of the Jurassic period.

Limestone A sedimentary rock made of calcium carbonate, usually made of particles from the skeletons of dead organisms.

Lode A vein or sheet-like structure of ore minerals which cuts across other rocks.

Magma Molten rock. If it reaches the surface and is extruded by a volcano, it is called lava.

Mantle The layer of the Earth between the crust and the core. It is rich in silica, iron and magnesium.

Marble To a geologist, marble is a metamorphosed limestone, made of calcium carbonate with a sugary texture in a variety of shades. But the quarryman or mason will give the name marble to any ornamental rock which wi'l take a polish.

Metamorphic rocks Those with a texture and appearance which have been changed by heat or by a combination of heat and pressure.

Mid-ocean ridge A line of submarine mountains running along the middle of each of the world's oceans, caused by volcanic action as convection currents in the mantle rise towards the crust.

Modified Mercalli scale A measure of the *intensity* of an earthquake - that is, the effect on people and property.

Moraine Material, mainly clay and boulders, left behind after the melting of a glacier.

Oil trap Rock structure, often a fault or a fold, in which a reservoir of oil may be found.

Opencast A method of extracting coal or minerals by excavating a pit rather than sinking mine shafts.

Orogeny An episode of mountain building, caused by two of the Earth's plates pushing into one another.

Palaeontology The study of fossils.

Plate The crust of the Earth, together with the topmost layer of the mantle, is divided into irregularly shaped 'jigsaw pieces' or plates.

Quartz Silicon dioxide. One of the commonest minerals in the crust, it is found in sedimentary igneous and metamorphic rocks, in a range of colours.

Richter scale A measure of earthquake *magnitude* – the amount of energy released by the 'quake.

Sea floor spreading The process by which the oceans grow, from the middle, at a rate of a few centimetres a year.

Seat earth A layer of fossil soil found underneath a coal seam.

Sedimentary rocks Those formed from sediments made up of fragments of older rocks, or the remains of dead organisms.

Seismogram The paper output from a seismograph, the instrument which records the shockwaves of earthquakes.

Subduction zone The region where an ocean plate, carried by a convection current, dives into the mantle. These zones are characterised by deep earthquakes and by volcanoes at the surface.

Terrane A chunk of crust which is a constituent part of a continent.

Trilobites Extinct marine arthropods whose bodies and carapaces were in three sections.

Vulcanism The activity of volcanoes.

CREDITS & INDEX

Figs 62, 63, 64, 65 Harry M. Parker, Peak District Mines Historical Society
Fig 67 Tarmac Building Materials Ltd
Fig 68 Peak Park Joint Planning Board
Figs 71, 73 British Coal Opencast
Figs 78, 79, 83, 87 Alan Bromley
Figs 84, 85 The Royal Cornwall Museum
Figs 88, 89, 90, 94 Camborne School of Mines
Fig 91 English China Clays
Figs 93, 118 Anna Grayson
Fig 96 ETSU, Department of Energy
Fig 97 G.A. Kellaway
Figs 101, 102, 104, 108, 111, 112 Bristol City Museum & Art Gallery
Fig 103 Arril Johnson Animation
Fig 105 Alaistair Hunter Photography
Fig 109 Chris Pellant
Fig 115 Ron Bailey
Fig 116 By permission of the Syndics of Cambridge University Library
Fig 117 Robert Stoneley, Imperial College
Figs 119, 120 BP
Fig 121 Swanage Town Council
Fig 123 Palaeo Services Watford
Fig 124 Martin Morgan Jones
Fig 125 London Underground Ltd
Fig 130 Museum of London
Fig 131 Crown copyright, Historic Royal Palaces
Fig 132 The Geologists' Association
Fig 134 Geoffrey Sherlock
Fig 135 Tony Timmington
Fig 137 Francis G. Dimes
All other pictures are from The Natural History Museum

ACKNOWLEDGEMENTS

I should like to thank the following earth scientists for their help and encouragement: Alun Ashton, John Bramley, Tony Brooks, Martin Brown, Desmond Clark, John Cope, Peter Crowther, Barry Dawson, Richard Edmonds, Dianne Edwards, Richard Fortey, Christopher Green, Stuart Monro, Roger Musson, Michael O'Shea, Eric Robinson, Ian Rolfe, Robin Shail, Peter Sheldon, David Sole, Chris Stringer, Ian Thomas, Alan Timms, Hugh Torrens, Susanna Van Rose, Ken Walton, David Webster, Stan Wood.

I should also like to thank Myra Givans, David Fogarty and Geoffrey Stalker for their editorial help; and colleagues at the BBC, especially my producer, Gordon Hutchings, for the hours of hard work developing Rock Solid from idea to product.

Designer: David Robinson
Artists: Gary Hincks, Sally Alexander
Printing and Binding in Singapore by Craft Print Pte Ltd
© Anna Grayson 1992

A catalogue record for this book is available from the British Library

ISBN 0-565-01108-1